Free Innovation

Free Innovation

Eric von Hippel

The MIT Press
Cambridge, Massachusetts
London, England

Set in Stone Sans and Stone Serif by Toppan Best-set Premedia Limited. Printed and bound in the United States of America.

Library of Congress Cataloging-in-Publication Data

Names: von Hippel, Eric.
Title: Free innovation / Eric von Hippel.
Description: Cambridge, MA : MIT Press, 2016. | Includes bibliographical references and index.
Identifiers: LCCN 2016009390 | ISBN 9780262035217 (hardcover : alk. paper)
Subjects: LCSH: Technological innovations--Economic aspects. | Inventors. | Innovations.
Classification: LCC HC79.T4 H557 1988 | DDC 338/.064--dc23 LC record available at https://lccn.loc.gov/2016009390

10 9 8 7 6 5 4 3 2 1

Dedicated to the free and user innovation research community.

Contents

Acknowledgments

It is very clear whom I must and wish to thank first—my editor and beloved wife Jessie. Jessie was very busy raising our children when I was working on my previous book. However, on this book she was able to resume a full editorial partnership with me. As an editor, Jessie is a joy to work with. She ranges effortlessly from advice on the basic structure of arguments to finding just the right publication or quotation to deepen understanding of an idea. This book is very much better because she applied her erudition and wonderful editing skills to improving it.

Next, I want to thank the colleagues who share research interests with me, and who join me on research projects and coauthored papers. In addition to the learning we all derive from the growing innovation research community around the world, each book I write builds upon a decade or more of collaborative research projects with close colleagues. In the case of this book, the research studies that were most central to both theorizing about and generating data regarding *Free Innovation* were conducted with Carliss Baldwin, Harold DeMonaco, Steven Flowers, Johann Füller, Alfonso Gambardella, Nikolaus Franke, Fred Gault, Christoph Hienerth, Jeroen P. J. de Jong, Youngbae Kim, Jari Kuusisto, Karim Lakhani, Susumu Ogawa, Pedro Oliveira, Christina Raasch, Ruth Stock, and Andrew Torrance. Thank you all so much for our wonderful collaborations and friendships! In large part because of our work together, I find research to be a very joyful profession, full of growth and color. I look forward to our next excellent adventures—which I hope will include even more frequent walks to local bakeries and coffee houses.

It is essential that research and theory building on free innovation accurately understand and build upon the very new ways innovation is being conducted in the digital age. For help with this essential aspect of the work, I have centrally depended upon learning from friends and colleagues who are themselves innovators at the leading edge of practice. Central among these are Jeff Davis and Jason Crusan at NASA, Chris DiBona at Google, Jim Euchner at Goodyear, Bernadette Piacek-Llanes at General Mills, and Venkatesh Prasad at Ford Motor Company. Thank you all as always, and I very much look forward to our next discussions and projects!

Of course, as the book took shape I needed to ask for thoughtful review of the central ideas. This can be painful for my expert colleagues, especially at early stages when ideas are evolving rapidly, and the manuscript is still rough. The greatest contributions to reviewing drafts of *Free Innovation*, and giving insights for important improvements, were made by my colleagues Carliss Baldwin, Yochai Benkler, Dietmar Harhoff, Joachim Henkel, and Andrew Torrance—thank you so much!!

Next, with respect to getting the book into print, I want to thank my long-time senior editor at the MIT Press, John Covell, and also Ellen Faran, former Director of the MIT Press. Between the three of us, we successfully negotiated an agreement stipulating that free digital copies of the book will be made available on the web at the same time as printed copies are placed on the market. "Free" is still a rare practice in academic book publishing, and the MIT Press, with its strong public service orientation, is a leader in thinking about and experimenting with the commercial feasibility of various approaches to open access. In the approach we are using with *Free Innovation*, sales of physical printed copies will compensate the MIT Press for the costs of producing the book. The simultaneous availability of free digital copies may increase sales of physical copies or reduce them—we really do not know. What we do know is that it is important for readers that at least some of the books and research materials they need can be obtained for free. I very much hope that ways will be found to make open access for academic books a more general practice in future years.

With respect to the beautiful cover design, I want to thank Yasuyo Iguchi, Design Manager at the MIT Press. She incorporated a research photo of a spark created by my father, Arthur von Hippel, in an especially lovely way.

Finally, I thank my two wonderful children, Christiana von Hippel and Eric James von Hippel. We interact a lot as a family, and we really do affect each other's thinking. I am grateful for the good ideas and encouragement they both have provided as I worked very hard on this book.

1 | Overview of Free Innovation

In this book I integrate new theory and new research findings into the framework of a "free innovation paradigm." Free innovation involves innovations developed and given away by consumers as a "free good," with resulting improvements in social welfare. It is an inherently simple, transaction-free, grassroots innovation process engaged in by tens of millions of people. As we will see, free innovation has very important economic impacts but, from the perspective of participants, it is fundamentally *not* about money.

I define a free innovation as a functionally novel product, service, or process that (1) was developed by consumers at private cost during their unpaid discretionary time (that is, no one paid them to do it) and (2) is not protected by its developers, and so is potentially acquirable by anyone without payment—for free. No compensated transactions take place in the development or in the diffusion of free innovations.

Consider the following example:

Jason Adams, a business-development executive by day and a molecular biologist by training, had never considered himself a hacker. That changed when he discovered an off-label way to monitor his 8-year-old daughter's blood-sugar levels from afar.

His daughter Ella has Type 1 diabetes and wears a glucose monitor made by Dexcom Inc. The device measures her blood sugar every five minutes and displays it on a nearby receiver the size of a pager, a huge advantage in helping monitor her blood sugar for spikes and potentially fatal drops. But it can't transmit the data to the Internet, which meant Mr. Adams never sent Ella to sleepovers for fear she could slip into a coma during the night.

Then Mr. Adams found NightScout, a system cobbled together by a constellation of software engineers, many with diabetic children, who were frustrated by the limitations of current technology. The open-source system they developed essentially hacks the Dexcom device and uploads its data to the Internet, which lets Mr. Adams see Ella's blood-sugar levels on his Pebble smartwatch wherever she is.

NightScout got its start in the Livonia, N.Y., home of John Costik, a software engineer at the Wegmans supermarket chain. In 2012, his son Evan was diagnosed with Type 1 diabetes at the age of four. The father of two bought a Dexcom continuous glucose monitoring system, which uses a hair's width sensor under the skin to measure blood-sugar levels. He was frustrated that he couldn't see Evan's numbers when he was at work. So he started fiddling around.

On May 14 last year, he tweeted a picture of his solution: a way to upload the Dexcom receiver's data to the Internet using his software, a $4 cable and an Android phone.

That tweet caught the eye of other engineers across the country. One was Lane Desborough, an engineer with a background in control systems for oil refineries and chemical plants whose son, 15, has diabetes. Mr. Desborough had designed a home-display system for glucose-monitor data and called it NightScout. But his system couldn't connect to the Internet, so it was merged with Mr. Costik's software to create the system used today.

Users stay in touch with each other and the developers via a Facebook group set up by Mr. Adams. It now has more than 6,800 members. The developers are making fixes as bugs arise and adding functions such as text-message alarms and access controls via updates. ... (Linebaugh 2014)

Free innovation is carried out in the "household sector" of national economies. In contrast to the business or government sectors, the household sector is the consuming population of the economy, in a word all of us, all consumers, "all resident households, with each household comprising one individual or a group of individuals" (OECD Guidelines 2013, 44). Household production entails the "production of goods and services by members of a household, for their own consumption, using their own capital and their own unpaid labor" (Ironmonger 2000, 3). Free innovation, therefore, is a form of household production.

How can individual consumers justify investing in the development of free innovations when no one pays them for either their labor or for their freely revealed innovation designs? As we will see, the answer is that free innovators in the household sector are *self-rewarded*. When they personally use their own innovations, they are self-rewarded by benefits they derive from that use (von Hippel 1988, 2005). When they benefit from such things as the fun and learning of developing their innovations, or the good feelings that come from altruism, they are also self-rewarded (Raasch and von Hippel 2013). (In chapter 11, I will compare the concepts of free innovation, user innovation, commons-based peer production, and open innovation. Each offers a lens able to bring different aspects of household sector innovation into sharp focus.)

The Nightscout project described above illustrates several types of self-reward. From the account given, we can see that many participants gain direct self-rewards from personal or family use of the innovation they helped develop. Probably many also gain other forms of highly motivating self-rewards, such as enjoyment and learning, and perhaps also strong altruistic satisfactions from freely giving away their project designs to help many diabetic children.

Due to its self-rewarding nature, free innovation does not require compensated transactions to reward consumers for the time and money they invest to develop their innovations. (Compensated transactions involve explicit, compensated exchanges of property—that is, giving someone specifically this in exchange for specifically that. See Tadelis and Williamson 2013; Baldwin 2008.) Free innovation therefore differs fundamentally from producer innovation, which has compensated transactions at its very core. Producers cannot profit from their private investments in innovation development unless they can protect their innovations from rivals and can sell copies at a profit via compensated transactions (Schumpeter 1934; Machlup and Penrose 1950; Teece 1986; Gallini and Scotchmer 2002).

Enabled by individuals' access to increasingly powerful design and communication tools, free innovation is steadily becoming both a stronger rival to and a stronger complement to producer innovation (Baldwin and von Hippel 2011). Even today, it is very significant in both scale and scope. In just six countries surveyed to date, tens of millions of individuals in the household sector have been found to collectively spend tens of billions of dollars in time and materials per year developing products for their own use (von Hippel, de Jong, and Flowers 2012; von Hippel, Ogawa, and de Jong 2011; de Jong, von Hippel, Gault, Kuusisto, and Raasch 2015; de Jong 2013; Kim 2015). Over 90 percent of these individuals met both of the criteria defining free innovation: (1) they developed their innovations during unpaid, discretionary time, and (2) they did not protect the designs they developed from adoption by others for free. The remainder were aspiring entrepreneurs within the household sector, motivated at least in part by the goal of selling their innovations.

Free innovation provides great value to household sector innovators in the form of the specific forms of self-rewards described earlier and also in the form of a general "human flourishing" associated with personal participation in innovation activities (Fisher 2010; Samuelson 2015). It also, as we will see, very generally increases both social welfare and producers' profits relative to a world in which only producers innovate (Gambardella, Raasch, and von Hippel 2016). For all these reasons, free innovation is well worth understanding better.

The Free Innovation and Producer Innovation Paradigms

Free innovation differs so fundamentally from producer innovation that the two cannot be incorporated in a single paradigm. In this section I therefore propose and describe a new free innovation paradigm and contrast it with the traditional Schumpeterian producer innovation paradigm. Figure 1.1 schematically depicts these two paradigms and the interactions between them. Each describes a portion of the innovation activity in national economies.

Generally, development activity in the free innovation paradigm is devoted to types of innovative products and services consumed by

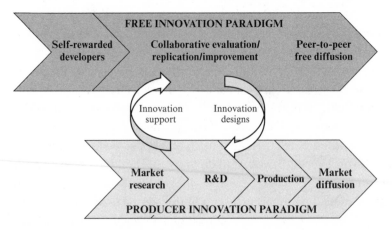

Figure 1.1
The free innovation paradigm and the producer innovation paradigm.

householders, not businesses. These represent a large fraction of Gross Domestic Product (GDP): In the United States and many other OECD countries, 60–70 percent of GDP is devoted to products and services intended for final consumption in the household sector (BEA 2016; OECD 2015). In contrast, innovation development activity in the producer innovation paradigm is devoted to addressing both consumer and industrial product and service needs.

As we will see, outputs from the two paradigms are complementary in some ways and competitive in others (Baldwin, Hienerth, and von Hippel 2006; Baldwin and von Hippel 2011; Gambardella, Raasch, and von Hippel 2016).

The free innovation paradigm

The free innovation paradigm is represented by the broad arrow shown in the top half of figure 1.1. At the left side of the arrow, we see consumers in the household sector spending their unpaid discretionary time developing new products and services. Discretionary time can be seen as "time spent free of obligation and necessity" (OECD 2009, 20), time devoted to activities that "we do not really have to do at all if we do not wish to" (Burda, Hamermesh, and Weil 2007, 1). Scholars have noted the potential value obtainable by producers and society when consumers increase the portion of discretionary time devoted to a range of productive uses (Von Ahn and Dabbish 2008; Shirky 2010). Innovation is clearly among such productive uses, as we will see in detail later.

As is implied by the position of the free innovation arrow in figure 1.1, which starts further to the left than the producer arrow, individuals or groups of innovators who have a personal use for an innovation with a novel function generally begin development work earlier than producers do—they are pioneers. This is because the extent of general demand for really novel products and services is initially often quite unclear. General demand is irrelevant to individual free innovators, who care only about their own needs and other forms of private self-reward that they understand firsthand. Producers, in contrast, care greatly about the extent and nature of potential markets and, as the rightward positioning of the producer arrow indicates, often wait for market information to emerge before beginning their own development efforts (Baldwin, Hienerth, and von Hippel 2006).

If there is interest in an innovation beyond the initial developer, some or many other individuals may contribute improvements to the initial design, as is shown at the center of the free innovation paradigm arrow. This pattern is visible in the Nightscout example presented earlier and is familiar in open source software development projects as well (Raymond 1999). Thus, in the Nightscout case, many individuals with an interest in helping children with Type 1 diabetes came forward to join the efforts of the project's initiators (Nightscout project 2016).

Finally, free diffusion of unprotected design information via peer-to-peer transfer to free riders may occur, as is shown at the right end of the free innovation paradigm arrow. (Free riders are those who benefit from an innovation but do not contribute to developing it. In that sense they get a "free ride.") Again, a pattern of diffusion to free riders is clearly visible in the Nightscout project.

Note that what is generally being revealed free for the taking by free innovators is design information, not free copies of physical products. In the case of products or services that themselves consist of information, such as software, a design for an innovation can be identical to the usable product itself. In the case of a physical product, such as a wrench or a car, what is being revealed is a design "recipe" that must be converted into a physical form before it can be used. In free peer-to-peer diffusion, this conversion is generally done by individual adopters—each adopter creates a physical implementation of a free design at private expense in order to use it. However, this is not a firm rule. Sometimes free innovators, motivated by altruism or other forms of self-reward, do create free physical copies of free designs to give to free riders. As an example, consider the worldwide e-Nable network. Founders of this network developed open source designs for inexpensive, 3D-printed artificial hands for children and adults who lack hands. Network members who own 3D printers donate their time to tailor the freely available hand designs to individual needs, and also donate the use of their personal printers to produce copies for free (Owen 2015).

The producer innovation paradigm

The long-established producer innovation paradigm centers on development and diffusion activities carried out by producers. The basic sequence of activities in that paradigm is shown on the lower arrow of figure 1.1. Moving from left to right on that arrow, we see profit-seeking firms first identifying a potentially profitable market opportunity by acquiring information on unfilled needs. They then invest in research and development to design a novel product or service responsive to that opportunity. Next, they produce the innovation and sell it on the market. In sharp contrast to household sector innovators, producers' innovation activities are *not* self-rewarding: the producer is rewarded by profit obtained via compensated transactions with others. (Of course, employees within firms may find their work personally self-rewarding. This can sometimes be reflected in their wages. In labor economics it has long been argued that firms can pay a lower wage as compensation for work that employees find more desirable in other ways. See Smith 1776, 111; Stern 2004.)

The producer innovation paradigm can be traced back to Joseph Schumpeter, who between 1912 and 1945 put forth a theory of innovation in which profit-seeking entrepreneurs and corporations played the central role. Schumpeter argued that "it is … the producer who as a rule initiates economic change, and consumers are educated by him if necessary" (1934, 65). The economic logic underlying this argument is that producers generally expect to distribute their costs of developing innovations over many consumers, each of whom purchases one or a few copies. Individual or collaborating free innovators, in contrast, depend only on their own in-house use of their innovation and other types of self-reward to justify their investments in innovation development. On the face of it, therefore, a producer serving many consumers can afford to invest more in developing an innovation than can any single free innovator, and so presumably can do a better job. By this logic, individuals in the household sector must simply be "consumers" who simply select among and purchase innovations that producers elect to create. After all, why would consumers innovate for themselves if producers can do it for them?

Schumpeter's views and the producer innovation paradigm came to be widely accepted by economists, business people, and policymakers, and that is still the case today. Sixty years later, Teece (1996, 193) echoed Schumpeter: "In market economies, the business firm is clearly the leading player in the development and commercialization of new products and processes." Similarly, Romer (1990, S74) viewed producer innovation as the norm in his model of endogenous growth: "The vast majority of designs result from the research and development activities of private, profit-maximizing firms." And Baumol (2002, 35) placed producer innovation at the center of his theory of oligopolistic competition: "In major sectors of US industry, innovation has increasingly grown in relative importance as an instrument used by firms to battle their competitors."

Details of the producer paradigm have changed over time. Significant producer innovations once were viewed as starting from advances in basic research (Bush 1945; Godin 2006). Later, studies of innovation histories showed that there often was not a clearly demarked research event initiating important innovations—although "technology first" innovations do exist and can be important (Sherwin and Isenson 1967). Still later, it was argued that research findings fed into all phases of innovation in what was called a "chain link" model of innovation (Kline and Rosenberg 1986). Today, many would argue that, while research inputs are indeed important, producers' innovation projects are more frequently triggered by discovery of unfilled needs. Hence the marketing mantra: "Find a need and fill it." In line with this view, current prescriptions for the management of innovation by producers generally follow the market-demand-initiated version of the producer paradigm shown in figure 1.1 (Urban and Hauser 1993; Ulrich and Eppinger 2016).

Finally, when contrasting the two paradigms, I note that the definition of free innovation differs from the "official" definition of producer innovation with respect to mode of diffusion. A free innovation is defined as one that diffuses for free, as I said at the start of this chapter. Within the OECD, in contrast, the definition for an innovation includable in government statistics requires that it be introduced onto the market: "A common feature of an innovation is that it must have been *implemented*. A new or improved product is implemented when it is introduced on the market" (*Oslo Manual* 2005, paragraph 150). (Note that the focus of both definitions is on *availability* for diffusion. There is no requirement that anyone actually adopt a free innovation that is available outside of the market or actually buy a producer innovation that has been introduced onto the market.)

In the Internet era, the OECD's producer-centric, definitional restriction that innovations must be "introduced on the market"—that is, made available for sale—is obsolete, I believe. Today it is also possible to make free innovations available for widespread diffusion independent of markets, often via the Internet. For example, the Nightscout innovations are widely diffused outside of markets via Internet-based free transfer. Open source software and open source hardware very generally are diffused in that same way. Excluding free innovations from government statistics via the present market-focused definition distorts our understanding of the innovation process. It will be important to update the OECD's definition, and there are calls to do this (Gault 2012).

Interactions between the paradigms
There are four important interactions between the free innovation paradigm and the producer innovation paradigm (Gambardella, Raasch, and von Hippel 2016).

First, identical or closely substituting innovation designs can be made available to potential adopters via both paradigms at the same time. For example, Apache open source Web server software is offered free peer to peer by the Apache development community *and* at the same time a close substitute is offered commercially by Microsoft. In such cases, peer-to-peer diffusion via the free innovation paradigm can *compete with* products and services that producers are selling on the market. The level of competition can be substantial. In the specific case just mentioned, 38 percent of Internet websites used Apache free Web server software in 2015. Microsoft was second, serving 28 percent of sites with its commercial server software (Netcraft.com 2015). Competition from substitutes diffused for free via peer-to-peer transfers can increase social welfare by forcing producers to lower prices. It can also drive producers to other forms of competitive responses with social value, such as improving quality or increasing investments in innovation development.

Second, innovations available for free via the free innovation paradigm can *complement* innovations diffused via the producer innovation paradigm. Free complements are very valuable to consumers as well as to producers. They enable producers to focus on selling commercially viable products, while free innovators fill in with designs for valuable or even essential complements. For example, a specialized mountain bike is of little value to a biker who has not learned specialized mountain biking techniques. Producers find it viable to produce and sell the specialized mountain bikes as commercial products, but largely rely on expert bikers innovating within the free paradigm to create and diffuse riding techniques as a free complement. That is, adopters generally learn new mountain biking techniques by a combination of self-practice and informal instruction freely given by more expert peers.

Third, we see from the vertical, downward-pointing arrow toward the right in figure 1.1 that a design developed by a free innovator may *spill over* to a producer and become the basis for a valuable commercial product. For example, the design of the mountain bike itself and many further improvements to it were developed by free innovator bikers. These designs were not protected by the free innovator developers, and were adopted for free by bike producing firms (Penning 1998; Buenstorf 2003). As we will see, adoption of free innovators' designs can greatly lower producers' in-house development costs (Baldwin, Hienerth, and von Hippel 2006; Franke and Shah 2003; Jeppesen and Frederiksen 2006; Lettl, Herstatt, and Gemuenden 2006).

Fourth and finally, we see from the vertical, upward-pointing arrow at the left of figure 1.1 that producers also supply valuable information and *support* to free innovators. For example, Valve Corporation, a video game development firm, offers Steam Workshop, a company-sponsored website designed to support innovation by gamers (Steam Workshop 2016). The site contains tools that make it easier for these individuals to develop their own game modifications and improvements and to share them with other players. Investments to support free design, such as the investment in Steam Workshop by Valve, can benefit producers by increasing the supply of commercially valuable designs that free innovators create (Gambardella, Raasch, and von Hippel 2016; Jeppesen and Frederiksen 2006; von Hippel and Finkelstein 1979).

The Need for a Free Innovation Paradigm

Thomas Kuhn defined scientific paradigms as "universally recognized scientific achievements that, for a time, provide model problems and solutions for a community of researchers" (1962, viii). Having a paradigm in place that is widely accepted, as in the case of the producer innovation paradigm, can be very helpful to scientific advancement. Once a paradigm is in place, as Kuhn writes, researchers can engage in very productive "normal science," testing and more precisely filling in pieces of a paradigm now assumed to be correct in broad outline. However, as Kuhn also explains, a paradigm never adequately explains "everything" within a field. In fact, observations that do not fit the reigning paradigm commonly emerge during the work of normal science, but are often ignored in favor of pursuing productive advance within the paradigm.

In the case of innovation research, empirical evidence related to free innovation in the household sector has been increasing during recent years. However, innovations developed and diffused without compensated transactions are entirely outside the Schumpeterian producer innovation paradigm—and, indeed, entirely outside the transaction-based framework of economics in general. Ignoring this evidence has allowed researchers to do productive work within the Schumpeterian paradigm, while deferring the work of incorporating free innovation into our paradigmatic understanding of innovation processes.

Eventually, Kuhn writes, conflicts between the predictions of a reigning paradigm and real-world observations may become so pervasive or so important that they can no longer be ignored, and at that point, the reigning paradigm may be challenged by a new one (Kuhn 1962). I propose that this situation has been reached in the case of transaction-free innovation processes developed and utilized by free innovators in the household sector. I therefore frame the free innovation paradigm both as a challenge to the Schumpeterian innovation paradigm, and also as a useful complement. Both paradigms describe important innovation processes, with the free paradigm codifying important phenomena in the household sector that the producer innovation paradigm does not incorporate.

With respect to my proposal of complementary innovation paradigms functioning in parallel, it is important to note that Kuhn developed his concept of paradigms to explain how revolutions in understanding occur in the natural sciences. Central to his argument was that a new paradigm replaces an existing one in a "scientific revolution." However, today the idea of paradigms has expanded beyond the study of natural sciences to the study of social sciences as well. In the social sciences, Kuhn's observation that new paradigms replace earlier ones is not always followed. Multiple paradigms may co-exist as complementary or competing perspectives. (See, e.g., Guba and Lincoln 1994.) It is with that view in mind that I propose the free innovation paradigm as a *complement* to the producer innovation paradigm rather than as a replacement. I am proposing that each usefully frames *a portion* of extant innovation activity.

Note that by proposing and describing the free innovation paradigm, I by no means claim that research needed to support it is complete. Indeed, I wish to claim precisely the opposite. A new paradigm is most useful when understandings of newly observed phenomena are emer-gent and when ideas regarding a possible underlying unifying structure are needed to help guide the new research (Kuhn 1962). This is the role I hope the free innovation paradigm described in this book will play. If it is successful, it will usefully frame and support important research questions and findings not encompassed by the existing Schumpeterian producer-centered paradigm, and so provide an improved platform for further advances in innovation research, policymaking, and practice.

In the remainder of this chapter, I give very brief overviews of the contents of the succeeding chapters. In chapters 2–7, I present and dis-cuss the core of the free innovation paradigm theory and related empir-ical findings. In chapters 8–10, I explore important contextual matters, including the broad scope of free innovation, the personal characteris-tics associated with free innovators' success, and the legal rights avail-able to free innovators. Finally, in chapter 11, I suggest and discuss some next steps for theory building, policymaking, and practice related to the free innovation paradigm.

Evidence for Free Innovation (Chapter 2)

The importance of free innovation depends in large part on its scale and scope. In chapter 2, we will see from national surveys that free innovation is important on both of these dimensions. In just six coun-tries surveyed to date, tens of millions of individuals have been found to collectively spend tens of billions of dollars on a wide range of products for personal use. A cluster analysis shows that about 90 percent of household sector innovators meet the two criteria specified in the definition of free innovation. Less than 10 percent of household sector innovators are interested in becoming entrepreneurs or in selling their innovations to producers.

A central feature of the free innovation paradigm is that it is free from compensated transactions. I explain what compensated transactions are, and how free innovators can viably innovate and freely reveal their innovations without resorting to them.

Viability Zones for Free Innovation (Chapter 3)

Innovation opportunities are "viable" for free innovators or producers only when their innovation-related benefits equal or exceed their innovation-related costs. In chapter 3, I adapt modeling discussed by Baldwin and von Hippel (2011) to describe the conditions required for innovation viability within three innovation "modes": free innovation by single individuals in the household sector of the economy, collaborative free innovation by multiple household sector participants, and innovation by producers.

Baldwin and I argue that the number of innovation opportunities that are viable for individual and collaborative free innovation is increasing rapidly as powerful, easy-to-use design and communication technologies become steadily cheaper. Across many fields, radical reductions in design costs are being driven by advances in computerized design tools suitable for personal use. At the same time, radical reductions in personal communication costs are being driven by advances in the technical capabilities of the Internet. Field-specific tools are following the same trend. For example, inexpensive and easy-to-use tools for genome modification have greatly increased the number of opportunities for biological innovation that are viable for free innovators in the household sector.

Pioneering by Free Innovators (Chapter 4)

As has already been discussed, the incentives and behaviors of innovators acting within the free innovation paradigm differ fundamentally from those of innovators acting within the producer innovation paradigm. As a consequence, innovation outcomes created within the two paradigms should also systematically differ. Indeed, identifying and clarifying such differences is a major value the free innovation paradigm can provide. In chapter 4, I illustrate this important point by showing that there are basic differences in the types of innovations developed, and in the timing of innovations developed, within the two paradigms. Free innovators, being self-rewarding, are free to follow their own interests. Unlike producers, they need not work only on projects they expect the market to reward. They therefore generally pioneer functionally new applications and markets prior to producers understanding the opportunity. Producer innovators generally enter later, after the nature and the commercial potential of markets have become clear (Riggs and von Hippel 1994; Baldwin, Hienerth, and von Hippel 2006).

Diffusion Shortfall in Free Innovation (Chapter 5)

In this chapter, I document and discuss an important difference between the free innovation paradigm and the producer innovation paradigm with respect to innovation diffusion. The difference springs from the fact that, unlike producers, free innovators do not protect their innovations from free adoption, and they do not sell them. As a result, benefits that free-riding adopters may gain are not systematically shared with free innovators—there is no market link between these parties. For this reason, free innovators may often have too little incentive, from the perspective of social welfare, to invest in *actively* diffusing their free innovations. In contrast, of course, producers do have a direct market link to consumers, so there should be no similar diffusion incentive shortfall within the producer innovation paradigm.

I review an initial empirical study that finds evidence compatible with diffusion incentive and diffusion investment shortfalls by free innovators (de Jong, von Hippel, Gault, Kuusisto, and Raasch 2015). I then suggest how to address a free innovation diffusion shortfall without resorting to the introduction of markets.

Division of Labor between Free Innovators and Producers (Chapter 6)

To this point in the book, we have seen that the free and producer paradigms systematically differ with respect to the innovators' incentives, activities, and outputs. Recall that the paradigms also interact. In chapter 6, I describe their major interactions and the effects of these in detail. Drawing upon modeling by Gambardella, Raasch, and von Hippel (2016), I explain that there is an opportunity for a division of innovation labor between free innovators and producer innovators that simultaneously enhances social welfare and producers' profits. Producers, my colleagues and I argue, will benefit by *not* investing in R&D that substitutes for innovations that free innovators develop. Instead, producers will—often but not always—benefit from investing in *supporting* free innovator design activities. Producers should then focus their own resources on development activities that free innovators do not engage in, such as refinements needed for commercialization. Social welfare, we find, will benefit from public policies that encourage producers to transition from a focus on in-house development to a division of innovation labor with free innovators.

Tightening the Loop between Free Innovators and Producers (Chapter 7)

As the value of free household sector design effort becomes clear, both free project sponsors and producers are increasing their efforts to "tighten the loop" between themselves and free innovators in order gain a more profitable fraction of that effort. Crowdsourcing calls by both free innovators and producers asking for assistance on innovation projects from the household sector are on the rise. Producers are also learning to support free innovators, seeking to channel their work into privately profitable directions.

The increased intensity of "mining" of household sector innovation resources by producers is likely to have both positive and negative effects on social welfare. On the positive side, projects that producers sponsor are likely to have commercial value, and so are likely to be commercially diffused when completed. On the negative side, producer creation and crowdsourcing of very attractive, "gamified" innovation project opportunities may draw free innovators away from innovation opportunities of perhaps higher social value, such as the pioneering innovations they might otherwise develop.

The Broad Scope of Free Innovation (Chapter 8)

In chapter 8, I document that free innovation extends well beyond product innovation—the type of innovation focused upon by almost all studies of household sector innovation to date. I do this by reviewing field-specific empirical studies by a number of colleagues that find significant levels of free innovation are present in services, processes, marketing methods, and new organizational methods.

The broad scope of free innovation development should not be a surprise. After all, the test for whether innovation opportunities are viable for free innovators has nothing to do with the specific nature of those opportunities. All that is required for opportunity viability is that free innovators' expected benefits exceed their costs.

Personality Traits of Successful Free Innovators (Chapter 9)

Nationally representative surveys find that from 1.5 percent to 6.1 percent of members of the household sector in six countries engage in product innovation. That is a lot of people: tens of millions. At the same time, it also means that at least 94 percent of householders are *not* engaging in product development. Since household sector innovation increases social welfare, and also generally increases producers' profits, it becomes useful to inquire about differences between householders who successfully innovate and those who do not. To that end, Stock, von Hippel, and Gillert (2016) explore personality traits significantly associated with successful household sector innovation at each of three major innovation process stages: having an idea for a new product or product improvement, developing a prototype implementing the idea, and diffusing the innovation to others. My colleagues and I find that successful completion of each successive innovation process stage is importantly affected by *different* factors. Building upon that information, we propose ways to increase innovation success rates in the household sector.

Preserving Free Innovators' Legal Rights (Chapter 10)

In this chapter, I review household sector innovators' legal rights to engage in innovation and innovation diffusion. Drawing upon work reported in Torrance and von Hippel (2015), I explain that free innovators have very strong legal rights, at least in the United States, with respect to both innovation development and innovation diffusion. Individuals are generally free to act however they choose as long as they do not materially harm others (Jefferson 1819; Chafee 1919). Individuals also have the fundamental right of free speech, which enables them to exchange information in order to work collaboratively and to diffuse their findings to others. Further, free innovators sometimes have important practical, legal, and regulatory advantages over producers.

Despite this generally favorable situation, free innovators' freedom to operate is frequently reduced, and free innovation costs raised, by regulations or legislation promulgated for other purposes—often without awareness that free innovation even exists. Torrance and I make specific suggestions for improvement, and also propose that it will be valuable to increase general social awareness of free innovation, and the benefits it brings to society.

Next Steps for Free Innovation Research and Practice (Chapter 11)

In chapter 11, I suggest several next steps in free innovation research, policymaking, and practice that I think will be valuable. I begin by setting expectations for the role the free innovation paradigm might usefully play in these new efforts. Next, I compare and contrast the research lenses offered by free innovation, user innovation, peer production, and open innovation, outlining question types for which I expect each lens to be especially useful. I then propose steps to improve the measurement of free innovation, a matter that is very important to further progress on research questions related to the free innovation paradigm. Next I suggest research steps useful to incorporate free innovation into innovation theory and policymaking. Finally, I suggest how the free innovation paradigm can help us to understand the economics of household sector creative activities even beyond innovation, such as "user-generated content" ranging from fan fiction to contributions to Wikipedia.

I conclude the book by again noting that free innovation, free from the need for compensated transactions and intellectual property rights, represents a robust, "grassroots" mode of innovation that differs fundamentally from the prevailing Schumpeterian model of producer-centered innovation. I suggest that the free innovation paradigm, presented and discussed in this book, will enable us to understand free innovation more clearly and apply it more effectively, with a resulting increase in social welfare and human flourishing.

In this chapter I present evidence that free innovation is a very substantial phenomenon with respect to the development of products consumed within the household sector. As we will see, today tens of millions of consumers annually spend tens of billions of dollars creating and modifying products to better serve their own needs. In fact, aggregate household sector product development expenditures rival the scale of business sector expenditures by producers developing products *for* consumers. Next, we will see that more than 90 percent of the developers of product innovations in the household sector meet both of the criteria for free innovation specified in chapter 1: the innovators develop their innovations during their unpaid, discretionary time; and they do not actively protect their designs from free adopters. The remainder are aspiring entrepreneurs. Finally, I explore the nature of transaction-free self-rewards central to the viability of free innovation, and discuss why it can make economic sense for free innovators to reveal their innovations for free.

Six National Studies

At the time of this writing, six national surveys have explored the scale and scope of household sector product innovation by product users. I begin with a very brief overview of the methods all these studies used. Full details will be found in the published reports on each. The six national surveys were carried out in the United Kingdom by von Hippel, de Jong, and Flowers (2012), in the United States and Japan by Ogawa and Pongtanalert (published in von Hippel, Ogawa, and de Jong 2011), in Finland by de Jong, von Hippel, Gault, Kuusisto, and Raasch (2015), in Canada by de Jong (2013), and in South Korea by Kim (2015). All six study samples included only new products and product modifications that had been developed by household sector individuals for personal or family use. To qualify for inclusion in our studies, we required that the developments provided useful functional improvements over products already available on the market, and that they had been developed within the three years prior to data collection. Aesthetic improvements were not included. Innovations that individuals developed at home for their jobs, rather than for personal or family use, were also not included in the study samples.

All six surveys utilized what are called nationally representative samples. A sample of this type is designed to mirror the demographic composition of a nation's population. For example, if a population contains a specific percentage of technically educated individuals, the sample will have a similar "representative" percentage of respondents with that characteristic. Because of this feature, we can project the findings derived from a nationally representative sample onto a nation's population at large. Data were collected by means of a questionnaire administered by telephone interviewers in the United Kingdom, Finland, and Canada and by means of Internet sites in the United States, Japan, and South Korea. Questions used in four of the six national studies were identical (UK, US, Japan, and South Korea). New questions were added to the fifth and sixth studies (Finland and Canada) to address additional issues. The full questionnaire used in the most recent study (Finland) is published in appendix 1 and also in de Jong (2016).

The scale of product innovation in the household sector

Recall that in the producer innovation paradigm, consumers are not expected to innovate—they are expected to consume. However, quite contrary to this conventional assumption, the data my colleagues and I collected found that 24.4 million people had developed or modified products for their own use in just the six countries surveyed to date (table 2.1). This quite large number is likely to be a very conservative measure of total household sector innovation development activities. As I noted above, the national six surveys included only *product* innovations developed for personal or family use. Service and process development activities in the household sector were not included, and are likely to also be of significant scale when measured.

Table 2.1
Fraction of individuals developing products for their own use in six countries.

	UK (n = 1,173)	US (n = 1,992)	Japan (n = 2,000)	Finland (n = 993)	Canada (n = 2,021)	S. Korea (n = 10,821)
Percentage of consumer innovators in the population aged 18 and over[a]	6.1%	5.2%	3.7%	5.4%[b]	5.6%	1.5%
Number of consumer innovators aged 18 and over[a]	2.9 million	16.0 million	4.7 million	0.17 million[b]	1.6 million	0.54 million

a. In all six surveys individuals under age 18 were excluded due to youth privacy considerations.
b. In Finland, the age range was 18–65.

The scope of consumer product innovation

The products developed by consumers addressed a wide range of household sector activities (table 2.2). Areas showing high levels of innovation mapped well upon major categories of unpaid time activities reported by consumers. For example, in the United Kingdom, sports, gardening, household chores, caring for children, and using computers were significant activities (Lader, Short, and Gershuny 2006).

A few brief descriptions of innovations reported by respondents for each of the innovation categories listed in table 2.2 will illustrate both the nature and the broad scope of product development by consumers (table 2.3).

Spending on Innovation Projects

In the household sector, individual projects typically were developed using relatively modest, "person-sized" expenditures. As can be seen in table 2.4, spending by individual innovators on their most recent projects in the six countries averaged from a few hundred dollars to a little

Table 2.2
Scope of product development by household sector users in various innovation categories.

	UK[a]	Japan[b]	US[b]	Finland[c]	Canada[d]	S. Korea[e]
Craft and shop tools	23.0%	8.4%	12.3%	20%	22%	16.4%
Sports and hobby	20.0%	7.2%	14.9%	17%	18%	17.9%
Dwelling-related	16.0%	45.8%	25.4%	20%	19%	17.9%
Gardening-related	11.0%	6.0%	4.4%	na[f]	na	na
Child-related	10.0%	6.0%	6.1%	4%	10%	10.9%
Vehicle-related	8.0%	9.6%	7.0%	11%	10%	6.5%
Pet-related	3.0%	2.4%	7.0%	na	na	na
Medical	2.0%	2.4%	7.9%	7%	8%	5.5%
Computer software	na	na	na	6%	11%	na
Food and clothes	na	na	na	12%	na	na
Other	7.0%	12.0%	14.9%	3%	3%	23.9%

a. Source: von Hippel, de Jong, and Flowers 2012
b. Source: von Hippel, Ogawa, and de Jong 2011
c. Source: de Jong, von Hippel, Gault, Kuusisto, and Raasch 2015
d. Source: de Jong 2013
e. Source: Kim 2015

more than a thousand dollars in time and materials combined. (In these calculations, time was converted to a money equivalent by using the average per-hour wage rate in each nation surveyed.) The range of project expenditures by respondents was wide, varying from almost nothing—projects accomplished very quickly using only materials on hand—to levels much higher than average. Other research on other innovation samples finds that individuals who spent significantly more than average are likely to be lead users—individuals at the leading edge of important market trends having a strong need for their creations. Lead users are also more likely than average users to develop products with potential commercial value (von Hippel 1986; Urban and von Hippel 1988; Franke, von Hippel, and Schreier 2006; Hienerth, von Hippel, and Jensen 2014, table 3).

Table 2.3
Examples of household sector product innovations in various categories.

Craft and shop tools	I created a jig to make arrows. The jig holds the arrow in place and turns at the same time, so I can paint according to my own markings. Jigs available on the market do not rotate.
Sports and hobbies	I developed luminous paduk (go) stones, so that you can play the game in the dark. Compressed glossy material on the surface of the stones looks identical to normal ones, and the feeling is also similar.
Dwelling related	Due to the weather, I wanted my washing machine to spin only. I modified it by changing the way the timer worked to give a spin-only option. I bridged one of the circuits and inserted a switch. I used a GPS system that can be operated by computer and small tags to create a mechanism for immediately finding objects that have become lost in the house. I used a microwave oven to create a half-pressure rice cooker. I drilled holes in a plastic container and used a large rubber band and small board to adjust pressure within the container so that the resulting rice tasted as good as that cooked with other sources of heat.
Gardening related	I made a device for trimming the tops of trees. It's a fishing rod with a large metal hook at the end. This enables me to reach the top of the trees, bend them down, and cut them.
Child related	I colored the two halves of a clock dial with different colors, so a child can easily see which side is past the hour and which before the hour. I used it to teach my kids to tell the time. I created a cloth expansion panel to enable me to fasten my Winter coat while wearing a baby carrier underneath. Helps keep me and my baby warm. Adapts to all my conventional zippers.
Vehicle related	I installed a display on my car key remote controller for parking location positioning. When unable to remember where I parked in a large parking lot or a parking lot with several floors, it can help save time and the effort in finding my car.
Pet-related	My dog was having trouble eating. I used a flat piece of laminated wood and put an edge around it like a tray to stop her bowl from moving around the kitchen. It is a successful innovation.
Medical	My mother had a stroke and became unable to use her limbs. I created a coat that was easy for her to put on and take off while in a wheelchair. The areas under the sleeves were cut open so that the sleeves could be opened and closed with special tape.
Computer software related	I am colorblind. I developed an iPhone camera app that identifies the colors of objects in a scene, and codes them for easy recognition.

Small expenditures on individual projects add up to quite large amounts in aggregate, simply because so many householders innovate. In the case of the United Kingdom, the United States, and Japan surveys, my colleagues and I were able to estimate total annual expenditures on product development in the household sector. In those three countries only, the national surveys included a question asking respondents how many projects they had carried out per year. That information, together with the data we had on the costs of innovators' most recent projects and the total number of innovators in each nation, enabled us to make the calculations.

Table 2.4
Individual expenditures on most recent user innovation project

	UK	US	Japan	Finland	Canada	S. Korea
Time spent on most recent project (person-days)	4.8	14.7	7.3	2.6	6.7	5.9
Average materials expenditure on most recent project	£101	$1,065	$397	207€	$58 (Canadian)	$368

Source: von Hippel, Ogawa, and de Jong 2011, table 1. Total expenditures include out-of-pocket expenditures for the specific project plus time investment calculated at average wage rate for each country.

As can be seen in table 2.5, multiple billions of dollars are spent annually by household sector innovators in the United Kingdom, the United States, and Japan in aggregate. Interestingly, as can also be seen in table 2.5, this level of expenditure is not so different from annual expenditures by consumer goods firms on developing products *for* consumers in those countries (von Hippel, Ogawa, and de Jong 2011). Again, this is an indicator that product development by householders is an activity of substantial scale.

Table 2.5
Total individual innovation expenditures per year on products for own use.

	UK	US	Japan
Average number of projects per year	2.7	1.9	2.6
Estimated total expenditures[a] by consumer innovators on consumer product development per year	$5.2 billion	$20.2 billion	$5.8 billion
Estimated consumer product R&D expenditures funded by producers per year[b]	$3.6 billion	$62.0 billion	$43.4 billion

a. Total expenditures include out-of-pocket expenditures for the specific project plus time investment calculated at average wage rate for each nation.
b. Calculated from national input-output tables.
Source: von Hippel, Ogawa and de Jong 2011, table 1.

Single vs. Collaborative Innovation

Recall from chapter 1 that innovators may develop their innovations either as single individuals or collaboratively with others. In the six surveys, most individuals reported having developed their most recent innovations alone, and 10-28 percent reported having innovated collaboratively (table 2.6). As I will discuss in chapter 3, this pattern makes good economic sense. Collaborative development can produce major cost savings for participants in larger projects, where substantial costs are being shared. For relatively small projects, however, such as the typical household sector projects documented here, it can be more efficient to innovate alone, and in that way avoid the costs of coordinating development work with others.

Table 2.6
Modes of innovation.

	UK	US	Japan	Finland	Canada	S. Korea
Innovation by single individual	90%	89%	92%	72%	83%	72%
Collaborative innovation	10%	11%	8%	28%	17%	28%

Is It Free Innovation?

Recall from chapter 1 that I defined free innovation as having two characteristics. First, no one pays free innovators for their development work, they do it during their unpaid, discretionary time. Second, free innovation designs are not actively protected by their developers—they are potentially acquirable by anyone for free. From the data in the six national surveys, we can directly conclude that more than 90 percent of the innovators surveyed fulfill these two criteria. With respect to the first, all six surveys asked respondents whether they had developed their innovations during their unpaid, discretionary time, and included data only from individuals who said this was the case. With respect to the second criterion, all six surveys provided a list of possible means of preventing free adoption, ranging from secrecy to patenting, and asked innovating respondents whether they had used any of these to protect their innovations. As can be seen in table 2.7, efforts to protect innovations by secrecy or intellectual property rights of any kind were quite rare.

Table 2.7
Household sector innovations protected by intellectual property rights.

UK	US	Japan	Finland	Canada	S. Korea
1.9%	8.8%	0.0%	4.7%	2.8%	7.0%

Of course, a general absence of investment in protection could mean simply that efforts to protect were seen as impractically costly by household sector innovators (Baldwin 2008; Blaxill and Eckardt 2009; von Hippel 2005; Strandburg 2008). If that were the case, these innovators might wish they *could* protect their innovations, and would do so if a low-cost way (say, very inexpensive forms of patenting) were to become available. Such a situation would render free innovation a fragile phenomenon, at risk of vanishing if a cheaper way to protect innovations emerged.

To test this possibility, my colleagues and I asked participants in the Finland and Canadian national surveys about their *willingness* to reveal their innovations freely. In Finland, 84 percent said they were willing to freely reveal their innovations to at least some people. Of these, 44 percent were willing to reveal their innovations to anyone and everyone, and an additional 40 percent were willing to freely reveal their innovations selectively to friends and others in their personal networks (de Jong, von Hippel, Gault, Kuusisto, and Raasch 2015). In the Canadian study, de Jong (2013) found that overall willingness to freely reveal was also 88 percent, with 66 percent of respondents willing to freely reveal to everyone, and an additional 22 percent willing to freely reveal selectively to their networks. In both Finland and Canada, in other words, it appears that free revealing is not simply an artifact of high costs of protection—a large fraction of household sector innovators are *willing* to freely reveal their innovations to some or all.

The Nature of Household Sector Innovators' Motivations

Earlier, I reasoned that innovation project opportunities can be "viable" for free innovators—those with benefits exceeding costs—only if those innovators are self-rewarded. After all, by my definition no one pays free innovators to innovate, and no adopter pays them for their designs. To assess this matter, in the Finland national survey respondents were asked about the types *and the relative strength* of the motives that drove them to innovate. Specifically, they were asked to distribute 100 percent of their motivations across five specific types of rewards. In addition, they were offered an "other" option to list any additional types of rewards that were important in their case.

Four of the five types of rewards asked about were known to be important motivators for contributors to open source software projects (Hertel, Niedner, and Herrmann 2003; Lakhani and Wolf 2005): personal use of the innovation (von Hippel 2005; Stock, Oliveira, and von Hippel 2015); personal enjoyment of innovation development work (Hienerth 2006; Ogawa and Pongtanalert 2011; von Hippel, de Jong, and Flowers 2012), personal learning and skill improvement (Bin 2013; Hienerth 2006; Lakhani and Wolf 2005), and helping others (Kogut and Metiu 2001; Lakhani and von Hippel 2003; Ozinga 1999). The fifth type of motivation measure was "to sell / make money." This motive does not fit within the free innovation paradigm: It is the main motivation of innovators within the producer innovation paradigm, and was included for that reason.

In the Finland study only, my colleagues and I had collected data from a sample of 408 household sector innovators that included both those who said they had developed innovations primarily for their own use (the 176 individuals represented in the tables in this chapter that refer to the Finland study) *and* those who made no such claim but filled out the full questionnaire nonetheless. My colleague Jeroen de Jong and I subjected this larger sample to a cluster analysis in order to group innovators with similar motivational profiles together (Green 1977; Schaffer and Green 1998). A four-cluster solution was found which was both in line with theoretical considerations, and had good stability (Cohen's kappa = 0.80) (source: de Jong 2015).

In figure 2.1 I report the fraction of the overall sample falling within each of the four clusters and also the distribution of motive types within each. As can immediately be seen, innovators in the household sector are generally driven by a mixture of motives rather than one pure type. Indeed, it was rare to find someone who was driven only by a single motive.

Each cluster in the figure is labeled with the name of the most important type of private benefit expected by household sector innovators in that cluster. "Participators" (43 percent of all the household sector innovators in the sample) expected the largest fraction of their innovation-related benefits to come from the self-reward of enjoyment and learning from participating in the innovation process itself. "Users" (37 percent of the sample) expected their largest fraction of benefit to come from personal use of the innovation they had developed. "Helpers" (11 percent) were those whose strongest motivation in the list of five asked about was to innovate in order to help others—altruism. "Producers" (9 percent of the sample) were most strongly motivated by the prospect of sales.

Next note that four of the five motives asked about involved expectations of *self-rewards*—compensated transactions were not required to obtain them. That is, when individuals say they use an innovation they have developed, they are self-rewarded—no one else is required to reward them. Similarly, if free innovators enjoy or learn from the process of developing their innovations, they are

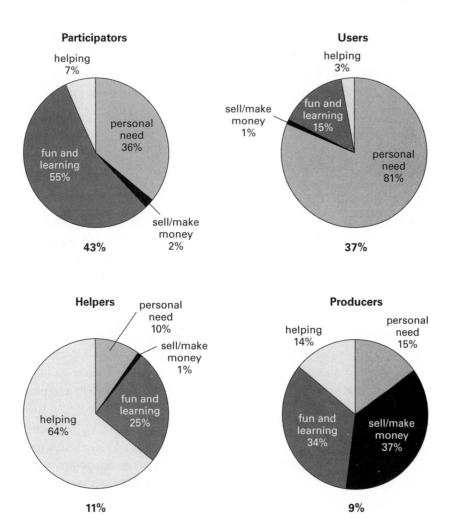

Figure 2.1
Household sector innovators in Finland clustered by mix of private benefits expected (*n* = 408).

self-rewarded—consuming those types of benefits also does not depend on transactions with others (Stock, von Hippel, and Gillert 2016; Stock, Oliveira, and von Hippel 2015; Raasch and von Hippel 2013; Franke and Schreier 2010; Hars and Ou 2002; Füller 2010). Also, as will be discussed further below, altruism too is a form of self-reward, not dependent upon compensated transactions. Only the last of the listed motives, the motive to "sell / make money," requires a compensated transaction with others.

From the findings shown in figure 2.1 we may conclude that innovators in three of the four clusters were free innovators—motivated by self-rewards almost entirely, and therefore finding it viable to invest in their innovation even if no one would pay them to obtain a copy. In sharp contrast, innovators in the producer cluster *were* significantly motived by the prospect of selling their creations. "Selling / making money" represented 37 percent of their total motivation. Of course, it is reasonable that a household sector survey will identify some individuals who are developing innovations to sell. Global Entrepreneurship Monitor (GEM) surveys find a similar fraction of individuals in the household sector (8.54 percent in advanced, "innovation driven" economies) to be in the early stages of entrepreneurial activity, with about half of these seeking to bring something novel to the market (Singer, Amorós, and Moska 2015, table A.3 and figure 2.14).

Individuals in the producer cluster differ from individuals in the three free innovator clusters with respect to behaviors as well as motivations. If one is seeking to sell, then developing designs likely to have value to many, investing in executing those designs very well, and protecting them from free adopters are all reasonable things to do. In line with these expectations, de Jong (private conversation, 2015) found that innovations developed by individuals in the producer cluster had significantly higher general value than innovations developed by individuals in the other three clusters. In addition, individuals in the producer cluster spent more developing their innovations (1,228 euros vs. an average of from 100 to 300 euros for the other three clusters). They also were also far more likely to protect their innovations via intellectual property rights (36 percent of innovators in the producer cluster did this vs. 3 percent or less in the other clusters).

Self-Rewards and Transaction-Free Activities

The concepts of self-reward and transaction-free activities, as I use them to describe the functioning of the free innovation paradigm, are connected: I define self-rewards as those private benefits that can be obtained without compensated transactions. A compensated transaction involves explicit or implicit arrangements to provide a specific party with "this" thing—perhaps a good, a service, or a financial instrument—in exchange for "that" thing. Therefore, when I say activities in the free innovation paradigm are transaction free, I mean that no compensated exchanges of this nature occur within it.

Compensated transactions are not involved when I gain personal use value from an innovation I develop and/or gain enjoyment and learning from engaging in the process of developing it. All of these reward types can be reaped without any requirement of related actions by or effects upon others—they are self-rewards. But what about the rewards associated with altruism that were asked about in the Finland survey? After all, others must adopt or benefit from my innovation before I can legitimately feel I have done something altruistic. Similarly, when I display or freely offer an innovation to others—reveal it without protections as the definition of free innovator requires—I may be hoping for a reward in the form of an increase in my personal reputation in the eyes of others (Lerner and Tirole 2002). In both of these instances, others have to do or experience something before I am rewarded. Why is this not a compensated transaction? The reason is that these hoped-for reactions are not an exchange of "this" for specifically "that" with a specific exchange partner. Instead, the freely revealed innovation is a casting of bread upon the waters, perhaps with expectations or hopes of return gifts in the form of "generalized reciprocity."

To clarify, let me digress briefly into the nature of gifts. First, note that compensated transactions, fitting the criterion of "specifically this for specifically that," can exist even without money or precise accounting as social transactions (Benkler 2006). Social transactions, Benkler explains, differ from economic transactions not in the absence of obligations for exchange, but in the precision of exchange. A market transaction has greater precision that "derives from the precision and formality of the medium of exchange—currency" (Benkler 2006, 109). In contrast, social exchanges are less precisely calculated. Benkler (ibid.) illustrates with a quote from Godelier in *The Enigma of the Gift*: "The mark of the gift between close friends and relatives ... is not the absence of obligations, it is the absence of 'calculation.'" Still, as Mauss (1966, xiv), quoting from the "Havamal" in his book *The Gift*, put it, "a gift always looks for recompense." Examining the elements of any gift,

Mauss discusses the three obligations involved—giving, receiving, and repaying—within which "the obligation of worthy return is imperative" (ibid., 41). As illustrated by Benkler, Godelier, and Mauss, when a gift is between specific known givers and recipients, it is a social transaction involving a compensated exchange of property, and such a gift is thus not "transaction free."

Second, note that gifts—such as those motivated by altruism in the case of our free innovators—*can* be transaction free when the givers expect generalized reciprocity rather than compensation from specific others. Generalized reciprocity, according to Sahlins (who first specified the term), is characterized by transactions that are generally accepted to be "altruistic," a "pure gift," and with expectations of recompense or direct material return being "unseemly" and at best "implicit" (Sahlins 1972, 193–194). It refers to "the return of a gift only indefinitely prescribed, the time and amount of reciprocation left contingent on the future needs of the original donor and abilities of the recipient; so the flow of goods may be unbalanced, or even one-way, for a very long period" (ibid., 279-280). Some have called generalized reciprocity "'helping a person backward,' where there is no chance of reciprocation by the person helped" (Ladd 1957, 291) or "paying it forward," described as the principle of "'I help you, and you help someone else'" (Baker and Bulkley 2014, 1493), but Sahlins expresses the essence, stating that "failure to reciprocate does not result in the giver of stuff to stop giving" (1972, 194).

Benjamin Franklin (1793, 178–179) made his important inventions available to all without patent protections. He explained his motives in terms of generalized reciprocity, saying "that, as we enjoy great advantages from the inventions of others, we should be glad of an opportunity to serve others by *any* invention of ours; and this we should do freely and generously." A smaller and more ordinary example of generalized reciprocity is the telling of the time to a stranger who stops one on the street to ask for that favor. You do not expect to see that individual again, nor to receive a return favor from precisely him or her. However, by contributing to a culture of generalized reciprocity, you can confidently expect that in the future some stranger will be willing to tell you the time if you ask. Very importantly, expectations of generalized reciprocity associated with a gift are transaction-free because, as was noted above, "failure to reciprocate does not result in the giver of stuff to stop giving" (Sahlins 1972, 194).

Within our context of free innovation, then, expectations of rewards in the form of generalized reciprocity such as gratitude or reputational enhancements evoked in the minds of others may motivate free innovators, and still not be compensated transactions made with specific others. However, there clearly can be a gray zone between transaction-free behaviors and transaction-based behaviors. For example, the number of active developers in an open source software development project may range from many to just a few. In the case of many developers, the situation contributors face might be most accurately called one of generalized reciprocity. However, as the number dwindles, awareness may grow that some specific person is developing and contributing X useful innovation to the commons and is doing so *because* another specific member will develop and contribute Y. In that case, the situation becomes one involving a compensated transaction.

To conclude, let me note that idea of transaction-free behaviors may seem odd, but in fact these are common in life—and justifiably so, in view of the costs and complexities that can be associated with arranging and executing compensated transactions (Tadelis and Williamson 2013). Baldwin (2008) points out that collaborative innovation projects, such as open source software development projects, are transaction free by design. She also points out that families and communities, when engaging in the activities of daily life, generally also engage in transaction-free interactions, often within a framework of generalized reciprocity. One can be confident, for example, that almost any adult will immediately rush to protect any young child from danger. According to Ladd (1957, 254), such help is offered "without the thought or expectations of reciprocation; and the reciprocation, when it does come, isn't considered a return but a new act of goodwill."

Discussion

Findings presented in this chapter show clearly that household sector innovation is significant in scale and scope. They also show that about 90 percent of household sector innovators fit the two criteria I have set for free innovation. That is, the innovators were motivated almost entirely by self-reward as compensation for their innovation-related investments, and they also did not protect their innovations from free-riding adopters.

In this section, I explain more richly *why* free innovators are willing to freely reveal their innovations. Although useful to us here, this topic has been explored in detail in previous work (e.g., Allen 1983; Harhoff 1996; Lerner and Tirole 2002; Harhoff, Henkel, and von Hippel 2003; von Hippel 2005, chapter 6). For this reason, I will provide only a brief summary of the main arguments.

The first fundamental point to note is that household sector innovators who are not rivals, and who do not plan to gain from having a monopoly on their innovations, generally do not lose anything by freely revealing their designs. For example, if I develop an innovation to help my diabetic child and have no interest in selling it, my own interests are in no way damaged if you adopt my design to help your diabetic child without paying me. This is true even if you did not contribute to the development work—that is, if you are a free rider. The same is true even if you are a producer who will make a great deal of money commercializing my free innovation, and who will not share any of the profits with me. After all, my self-reward—sufficient to induce me to develop the innovation—was to help my child. (Of course, there can be special reasons to restrict free revealing even in the case of non-rivalry. For example, free innovators who create medical devices that are complex or dangerous to use may freely reveal their designs only very selectively, wishing to avoid any health risk to adopters with lesser skills. See Lewis and Liebrand 2014.)

Second, given that one does not lose anything by free revealing, a passive absence of efforts to protect innovation-related information is the lowest-cost option for innovators. This is so because active exclusion requires investment to prevent revealing of design-related information that would otherwise leak out in the natural course of events (Benkler 2004; von Hippel 2005). For example, if you use an invention in public—say, if you ride an innovative bicycle in public—its design is to some extent "naturally self-revealing." That is, unless you invest in shrouding your bicycle's working parts, observers can to some extent understand its functioning via simple observation as you pass by (Strandburg 2008). Investments in protection can take the form of measures to maintain secrecy, as just described, and/or investments to prevent use of information that has been revealed via contracts or intellectual property rights.

Third, freely revealing rather than hiding an innovation can provide valuable, transaction-free rewards to free innovators well beyond the four types of self-rewards listed in earlier tables. For example, innovators who freely reveal their new designs may find that others then elect to improve their innovation, to mutual benefit (Allen 1983; Raymond 1999). Commercialization by producers also can create a source of supply for innovators that is cheaper than do-it-yourself production. For example, I might be pleased if a producer adopts my innovative medical device. Commercialization of my development would give me the convenience of buying copies I might need in future rather than having to make them for myself (Allen 1983). And, of course, revealing innovations for free can enhance innovators' reputations, sometimes leading to valuable personal outcomes like job offers (Lerner and Tirole 2002).

Despite the benefits of free revealing listed above, the option to protect one's innovation is open to all. Indeed, recall that many household sector innovators in the "producer" cluster in figure 2.1 do exactly that in pursuit of profit. Why do not more innovators, since the opportunity is open to all, opt for protection and commercialization instead of free revealing? A major reason, I surmise, is that even if an effort to commercialize might yield some profit in the end, investing time and money to realize that profit has opportunity costs associated with it. (An opportunity cost is the loss of potential gain from other alternatives when one alternative is chosen.) All household sector innovators—and all of us—have many things that compete for our time and attention. Household innovators in the producer cluster appear to have decided that commercialization is worth pursuing under their particular circumstances (Shah and Tripsas 2007; Halbinger 2016). In contrast, household sector innovators who choose the path of free innovation may simply prefer devote their time and money to following other opportunities.

3 | Viability Zones for Free Innovation

In this chapter I explore the conditions under which innovation *pays* for both free innovators and producers. Drawing heavily upon research carried out with Carliss Baldwin (Baldwin and von Hippel 2011), I first define and describe three basic innovation modes: free innovation by single individuals, collaborative free innovation by multiple individuals, and producer innovation. I then explore the conditions under which each of these modes is "viable"—that is, will provide a net benefit to innovators engaging in it.

Building upon innovation mode viability calculations, we will see that continuing improvements to free innovators' design tools and communication capabilities are making free innovation viable for an increasing range of innovation opportunities. As a result, it is reasonable to conclude that free innovation will steadily grow in importance relative to producer innovation.

Three Innovation Modes

The thinking and the analyses that I will describe in this chapter were first developed in a paper analyzing the viability of user and producer innovation modes (Baldwin and von Hippel 2011). In what follows, I apply this work, with slight modifications to definitions, to analyze the viability of free and producer innovation modes.

Recall from chapters 1 and 2 that free innovation involves innovations developed at private cost by individuals during their unpaid discretionary time and also involves innovation designs that are not protected by their developers and so are potentially acquirable by anyone "for free." Recall also from chapter 1 that two different modes of innovating occur within the free innovation paradigm: free innovation by single individuals, and free innovation by groups of collaborating individuals. Together with producer innovation, this gives us three basic "modes" of innovation:

• A *single free innovator* is an individual in the household sector of the economy who creates an innovation using unpaid discretionary time and does not protect his or her design from adoption by free riders.

• A *collaborative free innovation project* involves unpaid household sector contributors who share the work of generating a design for an innovation and do not protect their design from adoption by free riders.

• A *producer innovator* is a single, non-collaborating firm. Producers anticipate profiting from their design by selling it. It is assumed that, thanks to secrecy or intellectual property rights, a producer innovator has exclusive control over the innovation and so is a monopolist with respect to its design.

Viability of an Innovation Opportunity

A mode of innovation is *viable* with respect to a particular innovation opportunity if the innovator or each participant in a collaborative effort finds it worthwhile to incur the costs required to gain the anticipated value of the innovation (Arrow 1962; Simon 1981; Langlois 1986; Jensen and Meckling 1994; Scott 2001). This definition of viability is related to the contracting view of economic organizations (Alchian and Demsetz 1972; Demsetz 1988; Hart 1995), the concept of solvency in finance, and the concept of equilibrium in institutional game theory (Aoki 2001; Greif 2006).

In terms of benefits, we define the *value of an innovation*, denoted by v, as the benefit that a party expects to gain from converting an innovation opportunity into a new design—the recipe—and then turning the design into a useful product, process, or service. As was discussed in chapters 1 and 2, free innovators and producers benefit from innovations they develop in different ways. Free innovators benefit from self-rewards and do not protect their innovations from free adoption by others. Their self-rewards may include benefits from using the innovation, benefits from participating in the innovation process, such as fun and learning, and benefits from helping, such as the "warm glow" associated with altruism (Raasch and von Hippel 2013; Stock, Oliveira, and von Hippel 2015; Franke and Schreier 2010; Hars and Ou 2002). In sharp contrast, producers benefit from profitable sales, which may take the form of sales of intellectual property (a patent or license) or sales of products or services that embody the design. Ultimately, a producer's benefit derives from customers' willingness to pay for the innovative design.

With respect to *innovation-related costs*, the model of Baldwin and von Hippel (2011) includes four basic types:

• *Design cost*, *d*, is the cost of creating the design for an innovation. It includes the cost of specifying what the innovation is supposed to do. These instructions can be thought of as a "recipe" for the innovation that when implemented will bring the innovation into reality (Baldwin and Clark 2000, 2006a; Suh 1990; Winter 2010; Dosi and Nelson 2010).

• *Communication cost*, *c*, is the cost of transferring design-related information between project participants during the design process and of communicating design information to others to accomplish diffusion.

• *Production cost*, *u*, is the cost of carrying out the design instructions to produce the specified good or service. The inputs include the design instructions—the recipe—and the materials, energy, and human effort required to carry out those instructions. The output is the innovative product or service—the design converted into usable form.

• *Transaction cost*, *t*, is the cost of establishing property rights and engaging in compensated exchanges of property.

For any innovation opportunity, the condition for viability for any innovation process participant is straightforward: The value of the innovation to any individual or firm *i* (expressed as v_i), must be greater than the costs that innovator incurs in design, in communication with others, in production, and in transactions. That inequality is

$$v_i > d_i + c_i + u_i + t_i. \tag{1}$$

To simplify discussion of the viability of the three modes of innovation, Baldwin and I first focus only on design and communication costs. This allows visualizing the zones of viability for each innovation mode in two-axis charts. Later, when the bounds on viability for all three innovation modes with respect to design and communication costs have been established, we will reintroduce the other two dimensions of cost and show how they affect the results. Therefore, for now consider that, for a given innovation opportunity, a particular mode of innovation is viable if and only if, for each necessary contributor to that mode, design and communication costs are less than the value that contributor expects, i.e., that

$$v_i > d_i + c_i. \tag{2}$$

When is an innovation opportunity viable for a single free innovator?
Figure 3.1 illustrates the innovation opportunity viability zone for a single free innovator. Project design costs (d) are represented on the horizontal axis, project communication costs (c) on the vertical axis.

The pattern we see is simple but interesting. Recall that the effort of innovation is worthwhile for a single free innovator in the case of a specific design opportunity if v_i is greater than the individual's cost of design plus cost of communication: $v_i > d_i + c_i$. Recall also that Baldwin and I defined communication cost as the cost of transferring design-related information among project participants during the design process, or to accomplish diffusion.

Under this definition, communication cost is zero in the case of design development by a single free innovator because that individual "does not have to talk with anyone" to benefit from developing and using the innovation. For example, if I have the capability to develop a

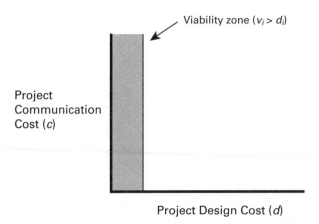

Figure 3.1
The viability region for an individual free innovator.

medical device or a type of sporting equipment to meet my own needs, I can "just do it," not communicating with anyone as I work on the project. I can then also use my improved equipment, again without the need to incur communication costs. In other words, our viability equation is reduced to $v_i > d_i$ in the case of design development and use by single free innovators. Because communication costs are zero, these individuals can find innovation development viable even if communication technology is very primitive, or if the costs of communication are very high for other reasons. This is why the shaded viability zone for single developers shown in figure 3.1 extends upward to include areas of high communication costs.

Note that single free innovators can choose to incur communication costs by investing in actively diffusing information about their innovation to potential adopters. However, they need not do this. Our definition of free innovation requires only that free innovators do not protect their design-related information—a choice that does not require investment in communication.

Even though single free innovators have no communication costs, they do have to expend time and money on design. An innovation project, therefore, will be viable for a single free innovator inside the vertically striped zone in figure 3.1, where $v_i > d_i$, but will not be viable outside that zone. That is, I would be willing to spend only up to d_i to respond to a specific innovation opportunity to improve my medical device, in view of the benefit of v_i that I expect. Of course, these values may be different for different individuals. If you need the same medical device somewhat more than I do, your v_i, and therefore your d_i, would be somewhat higher than mine.

When is an innovation opportunity viable for a collaborating free innovator?

Recall next that a collaborative innovation project is carried out by individuals who share the work. Open source hardware design projects, such as the Nightscout project described in chapter 1, and open source software projects are examples of collaborative innovation projects. In these projects, the participants are not rivals with respect to the innovative design they are creating. (If they were, they would not collaborate.)

Like single free innovators, collaborating free innovators need not invest in communicating with potential adopters. However, they must invest in communicating with others who are also contributing to the project. They must inform one another of ongoing design work, and they must coordinate to create a well-integrated full design. For this reason, communication costs in the case of collaborative free innovation projects are *not* zero, and so we are back to our viability inequality of $v_i > d_i + c_i$.

A collaborative innovation project offers two major advantages over innovation projects carried out by individual free innovators. The first major benefit from a participant's perspective has to do with output value obtained: Each individual participant incurs the design cost of doing a fraction of the project work but, if intending to use it, obtains the value of the entire design, including additions and improvements generated by others (von Hippel and von Krogh 2003; Baldwin and Clark 2006b). For example, if you and I have the shared goal of improving the design of a medical device used by diabetes patients, you may decide to design improvements to the electronics and I may decide to design improved hardware. In the end, if we both reveal our improvements, each of us gets to use the designs for *both* improvements while personally paying the design costs for only one of them.

Since designs are non-rival goods (both you and I can use a design at the same time—I am not competing with you for access), non-rival individuals considering creating an innovation should always prefer participating in a collaborative project to going it alone if a collaborative project is viable *and* the added costs of communication involved in a collaborative project do not exceed the savings individual participants gain by sharing design costs.

A second major advantage of collaborative projects over single innovator projects is that collaborative projects greatly expand the range of innovation opportunities that are viable for free innovators. This is because overall project costs are no longer limited to a level of design costs that are viable for a single individual.

Figure 3.2 shows how both of these factors play out. The horizontal extent (i.e., the width) of the rectangles in the shaded area at the bottom of the figure represents the viable amount of design costs for each individual participant in the collaborative project (d_i). (In the example

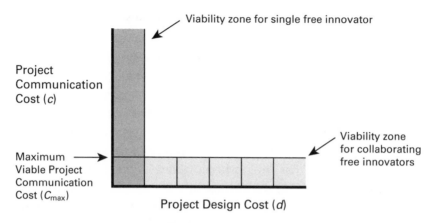

Viability zone for single free innovator

Project
Communication
Cost (*c*)

Viability zone
for collaborating
free innovators

Maximum
Viable Project
Communication
Cost (*C*max)

Project Design Cost (*d*)

Figure 3.2
Adding a viability region for a collaborative free innovation project.

given above, one rectangle might represent the effort of the contributor improving the diabetes device's electronics and another the effort of the person improving the hardware. The rectangles reflect the situation of individual contributors, and need not be of equal width.) The width of the shaded area across the bottom of the figure shows the scale of the design that can be undertaken by a collaborative free innovation project. As can be seen, the scale can be quite large— project costs can total to the aggregate willingness of many to pay for a portion of the design, and contribute their portion to the collaboration. If there are *N* contributors to a project, and each contributes his or her own part, the total design investment will be the sum of their individual design costs.

The top horizontal line in figure 3.2 (more specifically, the distance between that line and the horizontal axis, i.e., the height) represents the maximum viable communication costs for the project. It is calculated as the sum of the maximum communication cost that each contributor is prepared to bear, given the benefits they individually get from the collaboration. Conceptually, it should be clear that the lower the cost of communicating with the group, the lower the value threshold other members' contributions must meet to justify an attempt to collaborate. This means that low communication costs, as recently enabled by the Internet, are critical to the range of innovation opportunities for which the collaborative free innovation model is viable.

Lower communication costs affect the inequality $v_i > d_i + c_i$ in two ways. First, they decrease the direct cost of contributing, and so they increase the likelihood that an individual contributor will find joining the project and contributing to it worthwhile. Second, they increase the probability that others will contribute to the project. At a cost above C_{max}, demarked in figure 3.2, a collaborative project simply cannot get off the ground. But if communication costs are low for everyone, it is rational for each member of the group to contribute designs to the general pool and expect that others will contribute complementary designs or improve on his or her design. Again, this result hinges on the fact that the innovative design itself is a non-rival good. Each participant in a collaborative effort gets the value of the whole design, but incurs only a fraction of the design cost (Baldwin and Clark 2006b).

As makes economic sense, collaborative free innovation projects are generally "open" (that is, the innovation design information is freely revealed to all), because the cost of screening or other protective measures to exclude free riders would raise costs, and because free riders do not exert any negative effect on the free innovators. (Recall that free riders are those who benefit from a project design without making any contribution to it—they get a "free ride" when they adopt the innovation without paying or otherwise contributing.) Protective measures would shrink the pool of potential contributors, and so shrink the overall scale of the project. The network properties of the collaborative innovation model (the fact that the value to everyone increases as the total number of contributors increases) mean that this reduction in the contributor pool would reduce the value of the project to the contributors who remain as well as to free riders (Raymond 1999; Baldwin and Clark 2006b; Baldwin 2008).

Of course, any potential contributor might also decide to *not* develop and contribute an addition that could be viable for that contributor, hoping that someone else will do the work. This is the well-known incentive to free ride. But considerations such as urgency and self-rewards from performing the work can override such considerations for enough individuals to make a project viable.

When is an innovation opportunity viable for a producer?

Next, let us consider the space of innovation opportunities for which producer innovation is viable. Recall that a *producer innovator* is a single, non-collaborating firm that creates an innovation in order to sell it. Often producers can economically justify undertaking larger designs more easily than single individuals can, because they expect to spread their design costs over many purchasers.

Even though they are single organizations, producers, unlike single individuals, are affected by communication costs. They may use developers outside the firm, and then have to communicate with those outside individuals or organizations in order to coordinate. In addition, in order to justify investing in an innovation, they have to sell it. For that reason, they must invest in making potential buyers aware of what they have to sell via marketing communications. Such investments are often substantial, as the size of many producers' marketing budgets clearly attests.

Let us assume that a producer knows the development costs (d_p) and communication costs (c_p) that will be required to create the innovation and diffuse information about it to potential adopters. Let us also assume that the producer knows the value v_i that each potential adopter places on that innovation, as well as the number of potential adopters who would drop out of the producer's list of potential customers because they can self-supply more cheaply—in other words, that the producer knows each customer's willingness to pay for the producer's version of the innovative product or service. Following standard reasoning in microeconomics, the producer innovator can convert this knowledge about customers into a demand function, $Q(p)$, that relates each price it might charge to the number of units of the product or service it will be able to sell at that price. From the demand function, the producer innovator can solve for the price (p^*) and the quantity (Q^*) that maximize its expected revenues (net of production and transaction costs). Next, it can subtract its design (d_p) and communication (c_p) costs from this net revenue to calculate its expected maximum profit, P^*:

$$P^* = p^*Q^* - d_p - c_p. \tag{3}$$

If the producer anticipates positive profit for a specific innovation opportunity, then, as a rational actor, it will enter the market to supply the innovation. In other words, for that opportunity, the producer innovator model is viable. Conversely, if its anticipated profit is negative, the producer will not enter, and the producer model of innovation is not viable. As figure 3.3 shows, the zero profit line is a negative 45° line in the space of design and communication costs: $p^*Q^* = d_p + c_p$. For innovation opportunities within the triangle created by that line, the producer can expect profits. Those opportunities are therefore "viable" for the producer. Outside that triangle, innovation opportunities are not viable (Baldwin and von Hippel 2011).

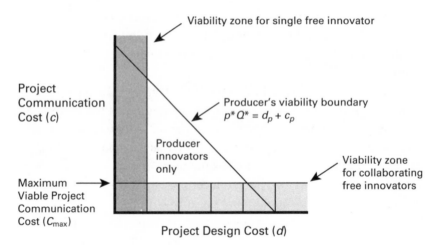

Figure 3.3
Adding a viability region for producer innovation.

Bringing Back Production Costs and Transaction Costs

Recall that at the beginning of this chapter, in order to focus on the contrasting effects of design and communication costs on the three modes of innovation, we made the simplifying assumption that production costs and transaction costs were similar across all three modes, and so had no effect on any mode's viability relative to the other two. I now bring these two costs back into consideration and discuss whether there are systematic differences in production or transaction costs across the three modes. In effect, we now will include all four cost variables—design costs d_i, communication costs c_i, production costs u_i, and transaction costs t_i—in our assessment of the viability of innovation opportunities:

$$v_i > d_i + c_i + u_i + t_i.$$

This discussion will show that production cost considerations may favor producers over free innovators in many cases today, but that production cost considerations are trending toward neutrality over time. Transaction cost considerations, on the other hand, favor free innovators over producers.

Production costs

Recall that a design is the *information* required to produce a novel product or service—the "recipe." For products that themselves consist of information, such as software, the production cost is simply the cost of making a copy of the design—essentially zero. For physical products, however, the design recipe must be converted into a physical form before it can be used. In such cases, the input consists of the design instructions—the recipe—plus the materials, energy, and human effort required to carry out those instructions. The output is a product—the design converted into usable form.

One of the major advantages producers have historically had over single free innovators and open collaborative innovation projects is economies of scale with respect to mass production technologies. Mass production, which became widespread in the early twentieth century, is a set of techniques whereby certain physical products can be turned out in very high volumes at very low unit cost (Chandler 1977; Hounshell 1984). The economies of scale in mass production generally depend on using a single design (or a small number of designs) over and over again. In classic mass production, changing designs interrupts the flow of products and incurs setup costs and switching costs, which reduce the overall efficiency of the process.

Can single free innovators or open collaborative innovation projects convert their various designs into physical products that will be economically competitive with the products of mass producers? Increasingly, the answer is Yes. Consider that today mass producers can design their production technologies to be independent of many of the specifics of the designs they produce. Such processes are said to provide "mass customization capabilities." Computer-controlled production machines can adjust to create a single unique item at a cost that is not different from producing a stream of identical items on those same machines (Pine 1993; Tseng and Piller 2003). When mass customization is possible, producers can, in principle, make their low-cost, high-throughput factories available for the production of designs created by single individuals and collaborative free innovation projects. Also, and increasingly, individuals can purchase production equipment designed for personal use such as personal 3D printers, and thereby have a low-cost production capability of their own that is entirely independent of the factories of commercial producers.

Of course, for a long time to come, there will continue to be instances in which the economies of mass production depend significantly upon careful and subtle co-design of products and product-specific production systems. In such instances, producer innovators will continue to have an advantage in designing and producing goods and services for mass markets.

Transaction costs

If producer innovators have a production cost advantage for some (but not all) production technologies, single and collaborative free innovators have an advantage with respect to costs for compensated transactions. By definition, they have none.

Consider that the ordinarily assumed transaction costs of innovation include the cost of establishing exclusive rights over the innovative design—for example, through secrecy or by obtaining a patent. Also included are the costs of protecting the design from theft—for example, by restricting access or by enforcing non-compete agreements (Teece 2000; Marx, Strumsky, and Fleming 2009). Finally, the costs of selling and receiving payment and the costs of protecting both sides against opportunism also are included in transaction costs and may be substantial. These may involve the cost of bargaining and writing contracts (Hart 1995), plus costs of accounting for transfers and compensation, and finally the costs of policing and enforcing agreements made (Williamson 1985).

Producer innovators *must* incur these transaction costs. By definition, they obtain revenue and resources from compensated exchanges with customers, employees, suppliers, and investors. A considerable amount of analysis in the fields of economics, management, and strategy considers how to minimize transaction costs by rearranging the boundaries of firms or the structure of products and processes. (For reviews of this literature, see Williamson 2000 and Lafontaine and Slade 2007.) For producer innovators, transaction costs are an inevitable cost of doing business.

Individual free innovators do not incur transaction costs. By definition they do not protect their innovation designs. Collaborative free innovation projects also do not sell products, nor do they pay members for their contributions. Transaction costs can creep in, of course, if individuals or groups decide not to fully relinquish claims on their intellectual property rights. For example, open source software projects generally assert a copyright over the software code created by their projects, doing so in order to preserve open access rather than limit it. The General Public License (GPL), based on copyright law, was explicitly designed to protect the rights of all to view, modify, and distribute open source software code bearing that license (Stallman 2002; O'Mahony 2003). The costs of enforcing the GPL are like classic transaction costs in that they assert and enforce property rights. Notwithstanding this minor exception, it is clear that free-revealing single free innovators and open collaborative innovation projects have a transaction cost advantage over producer innovators.

Hybrid Models of Innovation

Theory development is often best served by simplicity, such as in the three polar models of innovation Baldwin and I describe. In contrast, the world is often hybrid. A hybrid innovation model combines elements of the three polar models analyzed in previous sections of this chapter. Hybrids of the three basic models thrive in the real world. This is because the architecture of a design intended to achieve a certain function can often take a number of forms suited to development by combinations of our three basic models. For example, producers or free innovators can choose to modularize a product architecture into a mix of large components viable for development by producers only, plus many smaller components viable for development by single free innovators or open collaborative innovation projects (Baldwin and Clark 2000). As illustration, consider that Intel develops expensive and complex central processing unit (cpu) chips for computers, a design task that today may be viable for producers only. Complementary smaller software and hardware design opportunities are then viable for profit-seeking producers, and/or for free innovators, working singly or collaboratively.

Large indivisible design projects, which have traditionally been in the producer-only zone of figure 3.3, may become hybrids as a result of re-architecting and (often) modularization of traditional, producer-centered design approaches. For example, the costs of clinical trials of new drugs are commonly argued to be so high that only a producer innovator, buttressed by strong intellectual property protection for the drug to be tested, will find this development task viable. Increasingly, however, we are learning how to subdivide clinical trials—a large cost traditionally borne by drug producers—into elements suitable for voluntary, unpaid participation by collaborating individuals. This possibility has recently been illustrated in a trial of the effects of lithium on amyotrophic lateral sclerosis carried out by ALS patients themselves with the support of a website developed by the firm PatientsLikeMe (Wicks, Vaughan, Massagli, and Heywood 2011).

Discussion

Fundamentally, in a free economy, the organizational forms that survive are ones with benefits exceeding their costs (Fama and Jensen 1983a,b). Costs in turn are determined by technology and change over time. Chandler (1977) argued that the modern corporation became a viable form of organization (and the dominant form in some sectors) as a consequence of the decline in mass production costs due to technological advances, together with declines in transportation and energy costs. Adopting Chandler's logic, we should expect a particular organizational form to be prevalent when its technologically determined costs are low and to grow relative to other forms when its costs are declining relative to the costs of other forms.

To understand that the zones of viability for single and collaborative free innovation are growing over time requires only that one understand that design and communication costs for individuals have been decreasing due to exogenous technical trends, and that this is likely to continue.

Very generally, reductions in the cost of design in many fields are being driven by the rapidly declining cost and the increasing quality of personally accessible computer-based design tools. In fields where design is not implemented by digital methods, rapid progress in the development of field-specific tools is having the same effect. For example, in do-it-yourself biology, simple and powerful techniques to manipulate the genome are enabling individuals with little training to engage in genetic engineering and innovation (Delfanti 2012).

Reductions in communication costs for free innovation projects have been largely Internet-enabled. As in the case of design tools, "virtual reality" tools, and other new communication-related tools not yet envisioned, will extend the scale and scope of free innovation and diffusion. The central technological trend appears to be always toward increased fundamental understandings leading quickly or eventually to important capability advances accessible to household sector innovators.

With respect to production of physical products based on free designs, technical trends are increasingly empowering householders to complete the full development process by putting what they have designed into physical, usable form. As was mentioned earlier, personal and commercial production machines increasingly have the ability to produce a single unique item at a cost no higher than the cost per unit of a stream of identical items made with the same machines (Pine 1993; Tseng and Piller 2003).

In net, as a consequence of these exogenous technical trends, producer innovators—and innovation researchers and policymakers—increasingly must understand and contend with free single innovators and collaborative innovation projects as developers of innovative products, processes, and services (Benkler 2006, Baldwin and von Hippel 2011). To visualize the effect, imagine that figure 3.3 was populated with numerous points, each representing an innovation opportunity. As design and communication costs fall, each point moves down and to the left. Because of this general movement, some innovation opportunities would leave the region where only producer innovation is viable and cross into a region where single free and open collaborative innovation are also viable.

Although not all designs are equally affected, Baldwin and I believe that declining computation costs, communication costs, and single-unit production costs are having enough of an effect across the economy to change the relative importance of the three different models of innovation discussed in this chapter.

4 | Pioneering by Free Innovators

In chapter 1, I explained that the incentives and behaviors of innovators acting within the free innovation paradigm differ fundamentally from those of innovators acting within the producer innovation paradigm. As a consequence, innovation outcomes created within the two paradigms should also systematically differ. Indeed, identifying and clarifying such differences is a major value that the free innovation paradigm can provide to researchers, policymakers, and practitioners. In what follows, I illustrate this important matter with respect to innovation development. In chapter 5, I will do the same with respect to innovation diffusion.

The specific difference between paradigms that I will focus on is the pioneering role generally taken by free innovators in the case of new applications and markets, with producers following (Baldwin, Hienerth, and von Hippel 2006). I document this pattern and then explain changes in the rate of both free and producer innovation as a new field or application matures.

Why Free Innovators Pioneer

To understand the pioneering role of free innovators, recall from chapter 1 that producers generally expect to spread their design costs over many purchasers. However, to justify that expectation, producers need to be confident that many customers will in fact be interested in the product they plan to develop. They also need to be confident that they can somehow establish the monopoly rights needed to serve the market at a profitable price. In contrast, information about these things is irrelevant to individual free innovators. They care only about their own needs and their own self-rewards—matters that they understand firsthand.

Reliable information on the likely extent of demand generally does not exist at the beginnings of new applications and new markets where users are trying to do novel things—like experimenting with the first skateboards or with the first heart-lung machines. At that stage, markets are small and customers' needs are not clear. As a result, the information that a producer needs to determine whether acting on an innovation opportunity will be profitable is not available until well after the information that an individual innovator needs to determine the personal viability of that opportunity is available. This difference allows us to reason that free innovators will generally begin to innovate in new applications and new markets before producers do so (Baldwin, Hienerth, and von Hippel 2006).

Historical studies do support a pattern of free innovator pioneering. Many describe a sequence of events in which free innovator hobbyists enter new applications and markets ahead of producers in fields ranging from the development of the first aircraft (Meyer 2012), to the first personal computers (Levy 2010), and to the first personal 3D printers (de Bruijn 2010). Thus, Meyer documented that pioneering developers of the airplane were self-rewarding experimenters who freely shared their findings—free innovators—rather than early producers. "Early aeronautical experimenters were unusual, self-selected by their distinctive interest in the project of flight and their belief that they could contribute to it. They had an interest in the end goal. This helps explain why they would share their findings and innovations in clubs and journals and networks" (Meyer 2012, 7).

Pioneering by free innovators is also very visible in two quantitative studies that have explored the sources of innovation in new fields over time. I will briefly review the findings of those studies next.

Evidence of pioneering by free innovators in whitewater kayaking

The first of the two studies I will review involves innovation in equipment used in the sport of whitewater kayaking. Whitewater kayaking involves using specialized kayaks to maneuver in rough white water and also to perform acrobatic "moves" or "tricks" such as spins and flips. The sport began in about 1955 when a few adventurous kayakers began to develop methods of entering white water waves sideways or backward as a form of play. Soon, these "extreme paddlers" found one another and formed small communities to enjoy and develop the sport together. From those small beginnings, the sport of whitewater kayaking slowly grew to substantial size. In the mid 1970s there were only about 5,000 whitewater kayaking "enthusiasts" (frequent participants) in the United States (Taft 2001). By 2008 the sport had spread around the world and 1.2 million people were engaged in it, accounting for about 15 percent of all paddling activities (Outdoor Foundation 2009, 44). Expenditures by participants for gear and travel and other services reached hundreds of millions of dollars annually by 2009 (Outdoor Industry Foundation 2006; Outdoor Foundation 2009).

Hienerth (2006) and Hienerth, von Hippel, and Jensen (2014) studied the innovation history of whitewater kayaking from 1955 to 2010, carefully documenting the nature and source of innovations deemed "most important" by both expert kayakers and field historians. At the conclusion of this work, my colleagues and I had a sample of 108 important innovations that had been developed during four distinct phases in the sport's innovation history.

In phase 1 (1955–1973), whitewater kayaking was originated by adventurous kayakers as was noted earlier, and the basic outlines of the sport were laid down by the kayakers themselves. Kayakers were also the only developers of important equipment innovations in phase 1, collectively developing fifty. Near the middle of phase 1, small producers began to enter to serve the nascent market with commercial versions of kayaker-developed innovations. The producers developed no important innovations during that phase.

In phase 2 (1974–2000), whitewater kayaking techniques and equipment continued to develop rapidly. During phase 2, kayakers developed thirty important innovations and producers developed ten. Among the important producer innovations was the first rotationally molded plastic kayak hulls. These were much sturdier than the fiberglass versions that both kayakers and producers had been making previously. They were an essential enabler as kayakers steadily learned how to maneuver and play in increasingly rough water.

In phase 3 (1980–1990), which coincided with the middle years of phase 2, a few highly skilled kayakers, and eventually about a thousand, departed from the main practices of the sport to develop a novel form of whitewater play that they called "squirtboating." Squirtboating involved development of new maneuvers ("3D moves") that were carried out partially underwater in "squirtboats" of novel design. Squirtboats have very little buoyancy and were only safe in the hands of expert paddlers. Kayakers were the only innovators in phase 3, collectively developing ten important innovations.

In phase 4 (2000–2010), squirtboating largely merged back into the mainstream of the sport as a result of general adoption of the "rodeo kayak" hull design developed by kayakers. The hull of a rodeo kayak has high buoyancy at the center of the boat but very low buoyancy at the ends, and enables even non-expert playboaters to perform many 3D moves such as forcing the bow or stern of the boat underwater and doing end-to-end flips. In phase 4, kayakers developed no important equipment innovations and producers developed four.

The pattern and the sources of important whitewater kayaking innovations just described are summarized graphically in figure 4.1. As can be seen, kayak users clearly were the innovation pioneers in the new sport, preceding producers by more than 20 years. Further, kayak users were clearly the dominant source of important innovations in the sport. Of the 108 most important equipment innovations, 87 percent were developed by kayak users; only 13 percent were developed by all kayak producers in aggregate (Hienerth, von Hippel, and Jensen 2014). As can also be seen in the figure, the rates of important innovations by both users and producers decreased over time (a matter I will return to shortly).

The pattern in whitewater kayaking clearly fits the argument made at the start of the chapter. In line with the premise of that argument,

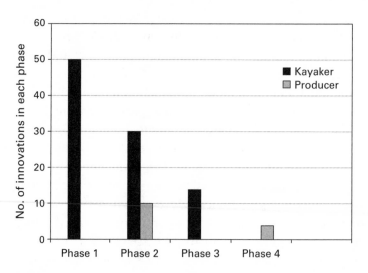

Figure 4.1
Source of important whitewater kayaking innovations over time. Source of data: Hienerth, von Hippel, and Jensen 2014, table 2.

innovating whitewater kayakers, when surveyed, reported being almost entirely motivated by self-rewards that could be obtained in full measure right from the start of the new sport. As can be seen in table 4.1, their self-reward came largely in the form of personal use of their kayaking innovations. They also freely shared their designs with peers and with producers (Baldwin, Hienerth, and von Hippel 2006; Hienerth, von Hippel, and Jensen 2014).

Table 4.1
Average motivations of household sector whitewater kayak equipment innovators.

Expected benefits from personal use	61%
Enjoyment from creating the innovation	17%
To help others (altruism)	10%
Learning from creating the innovation	8%
Other motives	2%
Potential profit from innovation sales	1%

Source: Hienerth, von Hippel, and Jensen 2014, table 6. Sample size: 201.

In contrast, producers are motivated by sales and profits. Clearly the small size of the potential market from the inception of the sport through the mid 1970s (there were only 5,000 enthusiast participants 20 years after the start of the sport, mostly designing and building boats to suit themselves) would have been less attractive to producers than was the large market of over a million participants that had emerged by 2010. Thus, in whitewater kayaking, the pattern of pioneering by kayakers—free user innovators—is clear and makes good economic sense.

Evidence of pioneering by scientists in scientific instruments
A second study shows the same clear pattern of user pioneering of new markets and applications. In this case the contrast is not between household sector free innovators and producers; it is between scientists employed by universities and firms and producers of scientific instruments. But the motivational distinction is the same: Scientists developed and improved novel instruments in order to use them in their scientific work, whereas producers developed novel instruments in order to sell them to many users.

William Riggs and I studied the sources and the timing of important innovations affecting two related types of instruments used in electron spectroscopy (Riggs and von Hippel 1994). Electron spectroscopy for chemical analysis (ESCA) and auger electron spectroscopy (AES) are both used to analyze the chemical compositions of solid surfaces (Riggs and Parker 1975; Joshi, Davis, and Palmberg 1975). In our 1994 study, Riggs and I identified 64 innovations judged to be important by both users and producers expert in these instrument types. The period of development studied began with the initial inventions in about 1953 and extended to 1983.

As can be seen in figure 4.2, the pattern of important innovations in ESCA and AES is very similar to that in whitewater kayaking. Scientists

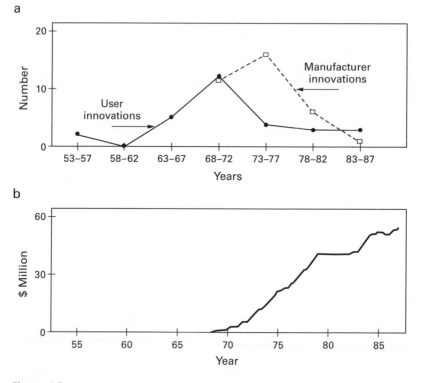

Figure 4.2
Source of important innovations in two types of scientific instruments over time. Graph a represents frequency of innovation; the first user innovations were developed around 1953, the first manufacturer innovations were commercialized around 1969. In graph b, the vertical axis represents millions of constant dollars, with a base period of 1982–84. Source: Riggs and von Hippel 1994, figure 2.

were the initial developers of both instrument types, and also of all early important improvements; producers only begin to innovate years later, with their first important innovations being commercialized in 1969. Also note that, just as in whitewater kayaking, the frequency with which both scientists and producers generated important innovations eventually declined, even though the combined sales of ESCA and AES instruments were rising (figure 4.2b).

The distinction between the innovation motives of scientists and producers is evidenced by a clear difference in the types of innovations they developed. As can be seen in table 4.2, scientists tended to develop innovations that enabled the instruments to do qualitatively new types of things for the first time. Such functions might have been of interest only to the innovators themselves, or they might also have been of interest to some additional fraction of the market. In contrast, manufacturers tended to develop innovations that made an instrument more convenient and more reliable in general—attributes of at least some interest to all potential customers. For example, scientist users were the first to modify the instruments to enable them to image and analyze magnetic domains at sub-microscopic scales, a capability of interest to only some users. In contrast, producers were the first to computerize instrument adjustments to improve ease of operation, a matter of interest to all users. Sensitivity, resolution, and accuracy improvements fall somewhere in the middle, as the data show. These types of improvements can be driven by scientists seeking to do specific new things with their instruments, or by producers applying their technical expertise to improve the products along known general dimensions of merit, such as accuracy (von Hippel 2005).

Table 4.2
Sources of scientific equipment innovations by nature of improvements effected

Type of improvement provided by innovation	Innovation developed by		Total (*n*)
	User	Producer	
New functional capability	82%	18%	17
Sensitivity, resolution, or accuracy improvement	48%	52%	23
Convenience or reliability improvement	13%	87%	24

Source: Riggs and von Hippel 1994, table 3. Sample size 64.

The difference in focus between scientist innovators and producer innovators can also be seen in the scientific vs. commercial importance of the innovations the two types of innovators developed. Riggs and I found that the scientific importance of scientist-developed innovations was on average significantly higher than that of producer-developed innovations ($p < .001$). However, the commercial importance of producer-developed innovations was on average significantly higher than that of scientist-developed innovations ($p < .01$).

How can we understand these patterns? I propose that the logic is identical to that discussed for the kayaking innovation study described earlier. Scientists, with their rewards based on the research value of the innovations to their own work and on the "scientific importance" of their developments, innovate first. They are not concerned with the potential size of a commercial market for their innovations. In contrast, producers wait until the nature, the scale, and the potential profitability of the market are clear before investing in innovation development. And when they do invest, producers tend to focus on developing innovations of interest to the entire market, such as convenience and reliability improvements, rather innovations of interest only to some segments of the market.

Why Do the Rates of Innovation by Both Free Innovators and Producer Innovators Decline?

We now understand why free innovators would be the ones to pioneer new markets and applications for use in the household sector. But what accounts for the decline in innovation frequency that is prominently visible in figures 4.1 and 4.2? As can be seen in the figures, this decline affects both user innovators and producer innovators, even as the markets increase in size. In contrast with pioneering, in other words, this effect applies to actors in both the free innovation paradigm and the producer innovation paradigm.

Baldwin, Hienerth, and I explained this pattern by arguing, first, that a new "design space" is opened up by the discovery of a new field or market (Baldwin, Hienerth, and von Hippel 2006). For example, the idea of intentionally engaging with rough white water in kayaks, as opposed to avoiding it as kayakers had historically done, was the creation of a new design space. This space contains all potential types of activities—not yet explored or even imagined at the time the new space is first perceived—that can be done in white water with manually powered kayaks. It also includes all possible designs of technique and equipment needed to realize them. However, any fixed design space has a limited number of valuable innovation opportunities within it. As time passes and search continues, it is reasonable that the valuable opportunities in the new design space will get progressively discovered and "mined out." The cost of searching for each of the increasingly rare undiscovered opportunities remaining will therefore rise, eventually making further searching unviable for innovators. This "mining out" is, my colleagues and I think, the reason why the number of innovations discovered in both whitewater kayaking and two types of scientific instruments declined with the passage of time.

Note, however, that in figures 4.1–4.2 we can see that the decline in the rate of producer innovation lags behind the decline in the rate of free or user innovation in both whitewater kayaking and scientific instruments. Why is this the case if the design spaces were in fact being mined out? The answer is that the steady rise in sales (shown for ESCA and AES instruments in figure 4.2b) made more and more innovation opportunities present in the design space financially viable for producers. Innovations of relatively smaller value to many people—the ones remaining as the space is mined out—can be justified only if there are many potential purchasers. In contrast, of course, the number of viable innovation opportunities in the design space does not increase for free innovators as the commercial market grows: their self-rewards are not affected by the size of the market.

Although "mining out" is a useful explanation for declining rates of innovation in our two cases, I caution that the effect exists only within a stable and even confining definition of a "legitimate" design space. For example, the rules of whitewater kayaking contests implicitly require that only manually propelled kayaks may be used—motors and motorboats are not allowed. If motors were allowed in the definition of the sport, the legitimate design space would clearly be larger, and "mining out" might take much longer. In the case of the two scientific instruments, the design space was defined to include only two instrument types with a common operating principle. If the design space had been widened to include *any* possible means to analyze the chemical composition of solid surfaces, it clearly would be much larger. Further, in cases in which there is no consensus on the boundaries of a design space (for example, today there is no apparent restriction on functions that people feel can be legitimately included in a smartphone), mining out is not a useful concept for understanding changes in the cost and rate of opportunity discovery and innovation development.

Finally, let me note that what was being mined out in our two case studies were opportunities for major innovations. Within any defined design space, opportunities also will exist at the level of what Hyysalo (2009) terms "micro innovations." These may never be mined out. For example, each whitewater kayaker is probably frequently motivated to make subtle adjustments to his or her equipment to better fit his or her physical condition, specific method of executing a given maneuver, and so forth as these change. Similarly, users of scientific instruments will continuously find needs for micro innovations to adapt to small changes in experimental protocols, changes to other instruments also being used in an experiment being done, and so on. Opportunities such as these may continue to exist within even a fixed design space and may be acted upon by both free innovators and producer innovators indefinitely.

Discussion

Recall that a basic distinction between the free innovation paradigm and the producer innovation paradigm is the absence of compensated transactions in the former: Due to its self-rewarding nature, free innovation is viable without either establishing intellectual property rights or selling to others via transactions. This inbuilt difference does not make one paradigm "better" than the other. It simply means that the types of innovations done within the two paradigms can differ systematically.

As we have seen in this chapter, one systematic difference is that free innovators tend to enter a new field early, pioneering new applications and markets by serving their own needs. Through their efforts, whether there is a potentially profitable commercial market becomes clearer. If the activities of the free innovators do reveal commercial potential, producers will respond by entering later, and then will focus their innovation development efforts on general needs such as convenience and reliability. If no potentially profitable market is revealed, free innovators will be the only ones to enter the field, and any diffusion will be via free peer-to-peer transfer only (Hyysalo and Usenyuk 2015).

The focus of free innovators on pioneering means that, at the time of development, free innovations will tend to be less "commercially important" in terms of immediate profits for producers than innovations producers develop (Riggs and von Hippel 1994; Arora, Cohen, and Walsh 2015). This is because, as we have seen in this chapter, producers enter later than free user innovators, when new markets are larger. For example, the commercial value of initial aircraft designs developed by free innovator hobbyists was essentially nil relative to the very large markets for present-day aircraft designs (Meyer 2012). However, the profit metric of innovation importance clearly must be seen in the larger context of free innovator pioneering. For a new development or market to *become* commercially important, it must first be pioneered—and here, as we have seen, free innovators play a very important role.

5 | Diffusion Shortfall in Free Innovation

In the preceding chapter I identified innovation pioneering as an important inbuilt difference between innovation development activities carried out within the free innovation paradigm and the producer innovation paradigm. In this chapter I identify an important inbuilt difference between the two paradigms with respect to innovation diffusion. By doing so, I further illustrate the research and practical utility provided by the free innovation paradigm.

The diffusion-related matter I will focus on is a systematic shortfall in free innovators' incentives to invest in diffusion of free innovations. I present evidence for that shortfall, and then argue that it is caused by the absence of a market link between free innovators and free-riding adopters. In a discussion at the end of the chapter, I suggest ways to address this situation.

"Market Failure" in the Free Innovation Paradigm

The value of free innovation to society comes in part from free innovators' satisfaction of their own needs via the innovations they develop. Social value is increased further if others also adopt and benefit from those same developments. Of course, to realize this second form of value, free innovations must diffuse from their developers to free adopters.

In chapter 2 we saw that more than 90 percent of innovators in the household sector do not attempt protect their designs from adoption by free-riding peers or producers. We also saw that most are quite willing to have their innovations diffuse for free to others. However, simply being willing to allow free riders to adopt a design if they wish to do so is by no means the same as *investing* to support diffusion to free-riding adopters.

Investment in diffusion by free innovators can increase social welfare because it is often the case that even relatively small investments can greatly reduce search and adoption costs for many free riders. For example, if I, as a free innovation developer, would invest just a little extra effort to document my open source software code more clearly, I could greatly reduce the time that perhaps thousands of adopters would require to install and use my novel code. Intuitively, it would seem that there would be a net increase in social welfare if I were to expend just that small extra effort.

To determine the optimal level of spending on diffusion in this case more exactly, it is useful to view the free innovation developer and the pool of potential free-riding adopters as a combined "system" for which we are seeking to maximize benefits. Assume that investments in diffusion of free innovations will lower adoption costs for free riders. Assume also that additional investments will lower adopters' costs at a declining rate. (For example, the first hour I spend improving my software code documentation might help clarify things a lot for free adopters, the second hour would contribute somewhat less additional clarity, and so on.) System benefit is then maximized at the point where an additional dollar of investment in diffusion by the free innovator—or anyone else in the system—reduces adoption costs by a dollar across all free adopters.

The question then is how to get to this optimal level of investment? The problem is that free innovators have to bear the costs of investments in diffusion, while free adopters get all of the benefits and do not share those costs. There is no market link that would enable a more appropriate allocation. Situations like these are described in economics under the heading of "market failure." With his evocative metaphor of an "invisible hand," Adam Smith described how pursuit of self-interest leads purchasers (whom he called "demanders") and producers jointly participating in a market to produce "always that precise quantity ... which may be sufficient to supply, and no more than that supply, that demand" (1776, 54, 56). A market fails when it does not get this balance right, and when the interaction of purchasers and producers fails to allocate resources efficiently (ibid., 55). Stated in present-day terms, a market failure exists when another possible outcome can make a market participant better off without making someone else worse off (Krugman and Wells 2006). Market failures, in turn, are regarded as a form of inefficiency, especially of information and resources, that calls for government intervention and remedy (Bator 1958; Cowen 1988).

The absence of a market link and resulting market failure affect only the free innovation paradigm. In the producer innovation paradigm, in contrast, there *is* a built-in direct market connection that rewards investments in diffusion. When customers buy a product for which the producer has monopoly rights, they transfer part of the benefit they derive from adopting the innovation to the producer in the form of a price higher than marginal cost. This gives the producer a monopoly profit that both motivates and rewards investments in diffusion to gain more sales. (However, as I will note at the end of the chapter, a different diffusion problem affects the producer innovation paradigm.)

The difference in levels of diffusion incentives within the two paradigms just described is not always so stark. It can be partially or even fully offset in cases where types of self-rewards that increase with diffusion are valued by free innovators. For example, presumably the self-rewarding "warm glow" of altruism experienced by a free innovator increases as the number of people who adopt his or her free innovation increases. One could say the same for self-rewarding pride of accomplishment. Reputation enhancements can also fit: at least sometimes, the greater the diffusion of an innovation, the greater the reputational gain for the developer.

Given all these factors, is there, in practice, a general shortfall in diffusion effort by free innovators? We do not have very good data on this matter yet but, as we will next see, the available evidence does point toward such a shortfall (de Jong, von Hippel, Gault, Kuusisto, and Raasch 2015; von Hippel, DeMonaco, and de Jong 2016).

Diffusion Performance of the Free Innovation Paradigm

There are two diffusion pathways a free innovation might follow. First, free information about an innovative design can flow directly from free innovators to peers, as figure 1.1 in chapter 1 shows. Second, as is also shown in that figure, design information can diffuse to producers for free, who then commercialize the design and sell it to adopters (Baldwin, Hienerth, and von Hippel 2006; Shah and Tripsas 2007).

In the six national representative surveys discussed in chapter 2, my colleagues and I collected data on innovation diffusion by both of these pathways. As can be seen in table 5.1, the rate of diffusion in these six countries by either pathway ranged from 5 percent to 21.2 percent of the innovation designs developed. On the face of it, this level of diffusion might seem low. However, in actuality, not all free innovations are candidates for diffusion. Recall that free innovators, motivated by self-rewards, may choose to create designs useful only to themselves. In those cases, an absence of diffusion is entirely appropriate. So, we must investigate further to see if the free innovation paradigm in fact underperforms with respect to diffusion.

Three Possible Manifestations of Diffusion-Related Market Failure

It seems to me that there is a sequence of three choices made by developers of free innovations that can each and collectively result in

Table 5.1
Development and diffusion of user innovations: results of national surveys. All studies sampled consumers aged 18 and over, with the exception of Finland (consumers aged 18 to 65).

| Source | Country | User innovators | | Innovations | |
		Percentage of population	Number	Diffused	Protected with IPRs[a]
von Hippel, de Jong, and Flowers 2012	UK	6.1%	2.9 million	17.1%	1.9%
von Hippel, Ogawa, and de Jong 2011	US	5.2%	16.0 million	6.1%	8.8%
von Hippel, Ogawa, and de Jong 2011	Japan	3.7%	4.7 million	5.0%	0.0%
de Jong et al. 2015	Finland	5.4%	0.17 million	18.8%	4.7%
de Jong 2013	Canada	5.6%	1.6 million	21.2%	2.8%
Kim 2015	S. Korea	1.5%	0.54 million	14.4%	7.0%

a. intellectual property rights

systematic diffusion shortfalls within the free innovation paradigm. First, free innovators may not elect to *design* an innovation of value to others. Second, even if a design does have general value, free innovators may not elect to invest in *development* to an extent justified by the total value of the design to themselves *and* free-riding adopters. Third, free innovators may not elect to invest in actively *diffusing* innovation-related information to reduce the adoption costs of free riders. I next discuss each of these three choices conceptually, drawing in the relatively small amount of data currently available.

Market failure type 1: Reduced general value of free innovators' developments

Even if slight modifications could make their designs serve others better, the incentives of self-rewarding free innovators may often be to focus only on their own needs. Of course, even if this is the path taken, the resulting free innovation might still be useful to others. It depends on how similar peoples' needs are with respect to that type of development. If you and I have the same needs, it will not matter if I develop a new product or service with only myself in mind—the product or service will turn out to be useful to you too. And, of course, if our needs are different, that will not be the case (Franke, Reisinger, and Hoppe 2009; Franke and von Hippel 2003).

The proportion of free innovators who do develop innovations potentially of benefit to others as well as to themselves must be determined empirically. Accordingly, my colleagues and I collected data on this matter via questions added to the Finland and Canada national surveys of household sector innovators discussed in chapter 2. In both surveys, respondents were asked questions to determine whether they thought that others would find their innovations valuable. Their responses were grouped into the three clusters shown in table 5.2. From the table, we see that, even without a market connection to free riders, 17 percent of the innovators thought their innovations would be of value to many others, and that an additional 30-40 percent thought their innovations would be of value to at least some others.

This fraction is likely a result of both needs held in common by free innovators and potential free-riding adopters, and self-rewarding motives that increase along with diffusion of the innovation. An

Table 5.2
General value of innovations developed by free innovators.

General value	Finland (n = 176)a	Canada (n = 1,028)
Cluster I: valuable to many or nearly all	17%	17%
Cluster II: valuable to some	44%	34%
Cluster III: valuable to few or to no one except the developer	39%	43%
Did not answer	0%	6%

Sources: For Finland, de Jong, von Hippel, Gault, Kuusisto, and Raasch 2015, table 5. For Canada, de Jong 2013, sections 3.3 a and b.

indication that the latter effect is playing a role comes from an analysis of data from Finland. Individuals who expressed *any* level of altruistic motivations (assigning at least one and at most 100 points to the innovation motive of helping others) were significantly more likely to have created a Cluster 1 innovation that could be of value to many than were individuals with no altruistic motivation at all ($\chi^2 = 9.2$, df = 2, $p = .01$) (de Jong 2015).

Market failure type 2: Suboptimal investment in design
Even if free innovators create a design of potential use to others, they may have no incentive to invest in improving the design to a level commensurate with potential value to themselves *and* free-riding adopters. For example, if fairly buggy code or roughly designed hardware will suit my personal needs, I may have no incentive to invest in refining my design, even if one thousand free riding adopters would benefit from my doing so. Free innovators will follow the viability calculations shown in chapter 3: They will invest in design only to the point that is optimal for themselves in the light of their particular constellation of self-rewards. Of course, when multiple free innovators collaborate on a project, design investment for the total project is likely to be higher than in single-developer projects.

Market failure type 3: Low diffusion effort by free innovators
The third possible manifestation of diffusion market failure is suboptimal investments to promote *diffusion* of a free innovation to free riders

who might benefit from it. In table 5.3 we see evidence compatible with this third type of market failure in the case of innovations developed by individuals in Finland (de Jong, von Hippel, Gault, Kuusisto, and Raasch 2015). In the data columns of that table we see that more than 75 percent of free innovators invested *no* effort in diffusion, even in the case of the Cluster 1 innovation designs that the developers thought had high general value. (Free innovators' self-assessments of general value may be right or wrong, but their efforts to diffuse for the benefit of others—the matter of interest here—will be a function of their own beliefs, not of the actual general value of their innovations.) Indeed, efforts to diffuse were so minimal that my colleagues and I had to use a very low threshold for our definition of active diffusion effort. Effort to diffuse an innovation peer-to-peer was deemed to exist if an innovator had simply shown the design to one or more peers. Effort to diffuse an innovation to a commercial firm was deemed to exist if an innovator had taken the initiative to show it to one or more commercial firms.

In addition to the finding of very low levels of diffusion effort in general, my colleagues and I found that in the case of peer-to-peer diffusion effort there is no significant relationship between diffusion effort exerted and the general value of the innovation ($\chi^2 = 2.5$, df = 2, $p = .285$). That is precisely what we would expect to see if there is a market failure of the type I am discussing here. Note that the pattern we see includes some free innovators making an effort to show innovations to peers that they themselves think have *no* general value (12 percent of cluster 3 in table 5.3). This can result if the free innovators have reasons to show their innovations for reasons not associated with

Table 5.3

Diffusion effort across clusters of general value in Finland.

Perceived general value	Diffusion effort made by free innovators	
	to inform peers	to inform producers
Cluster I: valuable to many	23%	19%
Cluster II: valuable to some	21%	6%
Cluster III: valuable to none	12%	0%

Source: de Jong, von Hippel, Gault, Kuusisto, and Raasch 2015, table 6.

general value. For example, they may wish to show a "cool project" to friends independent of whether they think those individuals would find it useful.

In contrast, free innovators' efforts to diffuse information about their innovations *to producers* were significantly related to their assessment of the general value of the innovations: The more generally valuable the developer thought an innovation was, the higher the likelihood that that individual would make an effort to inform producers about it ($\chi^2 = 12.2$, df = 2, $p = .002$). Of course, it is entirely reasonable that innovators will make an effort to inform producers only if they think the producer might find the innovation commercially interesting. After all, if there is no commercial value in the innovation, efforts to bring it to the attention of producers would be wasted. Still, despite this pattern, the fact that free innovators only informed producers about 19 percent of innovations they thought had the highest value to others (Cluster 1 in table 5.3) again suggests a market failure exists in the free innovation paradigm with respect to incentives to invest in diffusing free innovations to adopters.

Discussion

We now have a strong logical case and initial empirical support for the view that free innovators' investments in diffusion may generally fall short of the social optimum. As has been discussed, in the case of free innovation, this effect is due to a market failure "built into" the free innovation paradigm—the absence of a market connection between free innovators and free adopters (de Jong, von Hippel, Gault, Kuusisto, and Raasch 2015; von Hippel, DeMonaco, and de Jong 2016).

In this discussion I first note that the free innovation paradigm is not uniquely defective in this regard. There is a diffusion shortfall built into both the free innovation and producer innovation paradigms—but different adopter types are affected. Next I briefly consider three possible approaches to easing the diffusion incentive shortfall in the free innovation paradigm: a market solution, non-market solutions, and possible government policy solutions.

Exclusion of unskilled adopters

Diffusion shortfalls afflict *both* the free innovation paradigm and the producer innovation paradigm, but the causes are different. In the case of the free innovation paradigm, as we have seen, adoption costs are higher than the social optimum due to free innovators' "too-low" incentives to invest in diffusing them. In the case of the producer innovation paradigm, a diffusion shortfall results from producers' pricing above the marginal cost of production.

Consider that intellectual property rights enable producers to charge monopoly prices. (These rights are available to both free innovators and producer innovators, but only producers have a reason to want them: free innovators, giving their innovations away, have no interest in monopoly pricing.) Although monopoly pricing can increase producers' incentives to create innovations, they also create what is called "deadweight loss" with respect to the diffusion of innovations after they have been created. That is, monopoly prices exclude customers who would purchase the innovation and benefit from it if it were priced at the marginal cost of production, but who will not buy it at the higher prices set by the producer.

An interesting contrast can be made between the characteristics of the potential adopters denied access by these two different forms of adoption barriers. Those deterred from adopting a free innovation due to free innovators' inadequate investment in diffusion will tend to be relatively deficient in technical skills. In contrast, those deterred from adopting producer innovations by monopoly prices will tend to be those with less money. This pattern has not yet been studied, but my colleagues and I think it is both logical and clearly visible in everyday life. For example, people with technical skills do not need money to go to free Internet sites to "jailbreak" their smartphones and escape phone producers' restrictions. They can then download and use the latest free features. Millions in fact do this (Greenberg 2013). In contrast, people with money, and perhaps no technical skills, are more likely to pay phone producers' monopoly prices to buy the latest products equipped with the newest commercial features.

Solution via a market connection

As we have seen, a shortfall in the diffusion of free innovations can result from a lack of a market connection between free innovators and free-riding adopters. Accordingly, a straightforward solution could be to create a market connection between them. For example, one might devise some very cheap and easy form of intellectual property protection to induce free innovators to protect and sell their designs instead of giving them away. In other words, one could try to induce free innovators to elect to become producer innovators.

There is no doubt that this approach could work to some extent. As we saw in chapter 2, about 10 percent of household sector innovators already fall into the category of "producers" and behave in ways that would reward investments in diffusion. However, I myself do not consider it a preferred approach. Addressing a failure in the free innovation paradigm by inducing more household sector innovators to become producer innovators will also decrease the individual and social advantages that we have seen that free innovation provides. For example, it might reduce the scale of free innovators' pioneering of new applications and markets.

Non-market solutions

There are ways to work *within* the framework of the free innovation paradigm to increase the amount of diffusion of generally valuable free innovations. Two general approaches are: increase the strength of self-rewards that increase with diffusion, and/or lower the costs of creating and diffusing generally valuable innovations.

Interventions to increase self-rewards associated with diffusion generally alleviate all three manifestations of the diffusion market failure that are present in the free innovation paradigm today. This is because interest in creating a generally useful product is likely to be linked with interest in designing it well, and also in promoting widespread diffusion.

How can free innovators' self-rewards for investment in diffusion be increased? "Gamification" is one generally useful approach. It is known that games played without any practical output being obtained, like solitaire, are self-motivating activities (Fullerton 2008; Schell 2008; Gee 2003). Practical methods to manipulate and enhance such self-rewards are called gamification (Zicherman and Cunningham 2011).

Gamification strategies used to promote diffusion will vary by motive type. For example, one might increase levels of altruism-related self-rewards experienced by free innovators by providing them with better information on the number of adopters who would benefit by their investment in diffusion. An example of this strategy is the non-profit site Patient-Innovation.com (2016), which, among other activities, is working on collecting data on the most important needs of underserved medical patients with rare diseases (Oliveira, Zejnilovic, Canhão, and von Hippel 2015). The goal of the site's managers is to guide engineering classes and others seeking to contribute to projects valuable to these medical patients towards especially impactful opportunities. For free innovators motivated by reputation-related self-rewards, different gamification strategies would be useful. One might, for example, increase the likelihood of reputational gains for these individuals by publicly posting information about the admirable investments they have made to diffuse socially important free innovations.

With respect to lowering the costs of free innovation and diffusion, many specific costs seem reducible in many specific ways. For example, free innovators' costs of access to design and production tools can be reduced by support for "makerspace" communities, where access to costly tools is shared, and so rendered less expensive for individuals (Svensson and Hartmann 2016). Increased emphasis on open standards for design tools can lower the costs of acquiring and learning these tools, and also lower the cost of sharing design information created on a range of tools. Open sites for posting digital designs and design information can lessen the costs of diffusion for many—and so forth.

Diffusion of free innovations can also be increased by emphasizing support for collaborative free innovation projects over those carried out by single innovators. Available evidence shows that collaboratively developed designs diffuse much more frequently than designs created by single individuals. Thus, Ogawa and Pongtanalert (2013), who studied Japanese household sector product developers, found a rate of adoption by peers of 48.5 percent when the developers belonged to collaborative communities. When the developers did not belong to such communities, the adoption rate was sharply lower at 13.3 percent. Similarly, de Jong (2013) found, in a study of Canadian household sector innovators, that for collaborative projects the probability of peer-to-peer diffusion and adoption was 38 percent, whereas for single-innovator projects it was 20 percent.

I think there are two likely reasons for this effect. First, the needs addressed by collaborative projects are likely to be more general—after all, at least several collaborators are interested. Second, the information available to free adopters from collaborative innovation projects is likely to be much richer than that from single innovator projects. This is because participants in a collaborative project must document their activities to coordinate their work, something that single innovators need not do. This richer information, created for internal project use, can then costlessly spill over to the benefit of free adopters.

The case for governmental support

Some of the measures just described, such as support for need information sites, could benefit from governmental support. But why should government pay any attention to ameliorating a diffusion failure afflicting only the free innovation paradigm? Most fundamentally the answer is that, as will be explained in the next chapter, the diffusion and adoption of free innovation designs by those who benefit from them increases social welfare. With rare exceptions, such as the design of dangerous goods, society benefits if designs are public goods, available to anyone to use or study for free. Increased social welfare is, of course, the fundamental justification for governmental interventions in general (Machlup and Penrose 1950; Nelson 1959; Arrow 1962).

By analogy, governments today invest to cure and offset defects afflicting the producer innovation paradigm, notably by creating and supporting elaborate and very expensive intellectual property rights systems. They justify these investments and policies in terms of expected increases in social welfare. Investments in the free innovation paradigm under the same justification would only level the playing field.

6 | Division of Labor between Free Innovators and Producers

In this chapter, I explain the value of a division of innovative labor between free innovators and producer innovators. As Gambardella, Raasch, and von Hippel (2016) show, both social welfare and producer profits very generally increase if producers avoid developing types of innovations that free innovators already make available "for free." Instead, as my colleagues and I argue, producers should learn to focus on developing innovations that *complement* free innovation designs rather than substitute for them. Further, innovation tasks can and should increasingly be shifted to free innovators as their capabilities increase—that is, they should be shifted to what standard economic models think of as the demand side of markets.

I will begin by reviewing four basic interactions between the free and producer innovation paradigms. Then I will explain how my colleagues and I modeled the relationships among these interactions, and the effects that we found on both producers' profits and social welfare. As we will see, under some conditions producers can profit by actually subsidizing free innovation.

Four Major Interactions between the Paradigms

Recall from chapter 1 that there are four separate types of interaction between the free innovation paradigm and the producer innovation paradigm. These were represented schematically in figure 1.1, which for convenience is reproduced here as figure 6.1. First, designs diffused from peer to peer via the free innovation paradigm can *compete with* products diffused by producers via the market, resulting in what my colleagues and I call a *free-contested market*. Second, designs diffused from peer to peer via the free innovation paradigm can *complement* products and services diffused by producers via the market, a situation we call a *free-complemented market*. Third, as is indicated by the

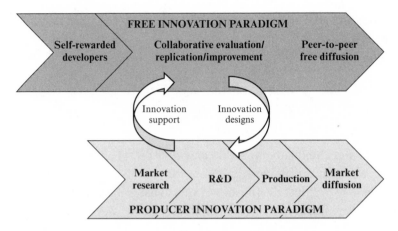

Figure 6.1
Interactions between the free innovation paradigm and the producer innovation paradigm. (Same as figure 1.1, reproduced here for the reader's convenience.)

downward-pointing arrow, free innovators "spill over" their free designs to free-riding start-up firms or incumbent producers. Fourth, as is indicated by the upward-pointing arrow, producers can supply tools and platforms to both support and shape free innovation.

In chapter 1, I briefly described the four paradigm interactions. Here I will describe what we know about each of them in more detail. This will provide a rich context for an exploration of producer strategies related to these interactions, and their effects on social welfare.

Free-contested markets

When an innovation is diffused for free to consumers via the free innovation paradigm and is a full or partial substitute for a product diffused by producers, producers face what Gambardella, Raasch, and von Hippel (2016) term a *free-contested market*. Free-contested markets involve a source of competition for producers that has not been contemplated in standard models of monopolistic or imperfect competition (see, e.g., Robinson 1933; Chamberlain 1962).

In a free-contested market, consumers as a group benefit from having access to the additional, non-market choice of free innovations and innovation designs. Some consequences of this situation have been studied in the case of competition among open source and closed source software suppliers (Casadesus-Masanell and Ghemawat 2006; Economides and Katsamakas 2006; Sen 2007). In that context, producers were found to lose profit from open source innovations distributed for free, even if those innovations were not full substitutes for producer commercial products. It was also found that consumers benefit from the existence of the open source alternative *unless* it forces proprietary firms to exit the market, leaving free, partial substitutes as consumers' only option. Loss of the producer option reduces the benefit to consumers because the two alternatives typically are not perfect substitutes—some consumers will prefer one and some the other (Kuan 2001; Baldwin and Clark 2006b; Casadesus-Masanell and Ghemawat 2006; Lin 2008).

Free-complemented markets

With respect to free-complemented markets, consider first that individual products or services are components within larger systems. For example, mountain bikes are a product that fits within a system of complements ranging from mountain biking techniques to helmets, tire pumps, navigation devices, and lights. From the perspective of the producer of any product or service within such a system, the other elements of the system are complements that range from useful to essential and that therefore add to the value of that "focal" product or service (the one I am focusing on). Thus, if I buy a specialized mountain bike and want to use it skillfully, I need the essential complement of mountain bike riding techniques. Biking techniques are largely diffused from peer to peer by free innovators rather than being sold. In other words, mountain bike producers are participating in and benefiting from a free-complemented market. The market for specialized mountain bikes would be much smaller without the free complement of mountain biking techniques.

Free-complemented markets can involve products that are separate from but complementary to producer products, as in the case of the mountain bike riding techniques just mentioned. They can also involve modifications or complements built onto or into producers' products or platforms. With respect to the latter, consider software modifications and additions that complement the value of basic commercial software products in fields ranging from music software to computer gaming software (Jeppesen and Frederiksen 2006; Prügl and Schreier 2006; Boudreau and Jeppesen 2015; Harhoff and Mayrhofer 2010). The evidence for the widespread presence of free-complemented markets runs counter to the conventional assumption that only producers provide complements, although customers are able to select and assemble them (Schilling 2000; Jacobides 2005; Adner and Kapoor 2010; Baldwin 2010).

In the case of systems of complements, producers may select the most commercially advantageous elements of a system to produce and sell. They will then prefer that the complements they do not sell will be provided to their customers in the form of free complements rather than as commercial products or services sold by other producers. The reason is that producer complementors seek to profit from the complements they provide, whereas free innovators do not. Free complementors therefore leave more profits available for the producer to extract from the system (Baldwin 2015; Baldwin and Henkel 2015; Henkel, Baldwin, and Shih 2013). For example, if free innovators provide the complement of biking techniques "for free," the value of the system of mountain bike plus mountain biking techniques to the mountain bike purchaser increases. A producer of mountain bikes that has monopoly power could extract some or all of the increased system value created by the free technique innovations by charging more for bikes.

Free spillovers of design information to producers

The third interaction between free and producer paradigms involves the spillover of free design information to producers (represented by the downward pointing arrow in figure 6.1). Producers can adopt free designs they think are likely to be profitable and commercialize them for the market at large. Research shows that such design spillovers can be highly valuable to producer firms, providing higher sales revenues, higher gross margins, and longer product life cycles for the producer (Lilien, Morrison, Searls, Sonnack, and von Hippel 2002; Winston Smith and Shah 2013; Franke, von Hippel, and Schreier 2006; Poetz and Schreier 2012); Nishikawa, Schreier, and Ogawa 2013).

Evidence for the importance of free designs to producers is illustrated by studies that explore the sources of all "important" innovations in a field. At the time of this writing I am aware of four empirical studies of this type that are focused on consumer products and services. Shah (2000) studied the source of important innovations in four sporting fields; Hienerth, von Hippel, and Jensen (2014) did the same in the specific sport of whitewater "playboat" kayaking; Oliveira and von Hippel (2011) studied the sources of important retail banking innovations; and van der Boor, Oliveira, and Veloso (2014) studied the sources of important innovations in mobile banking. These authors found that designs created by individual and collaborating users in the household sector accounted for a very significant fraction (from 45 to 79 percent) of all "important" innovations commercialized by producers in those fields. The innovative designs in the four studies were very rarely protected by their household sector developers: they were free innovations.

Cost savings for producers that adopt free designs can be estimated by first calculating producers' per-innovation design costs for innovations they *do* develop. That number can then be used to roughly estimate producer design cost savings in the case of each design adopted from free innovators. Data were collected for this calculation in the case of the whitewater kayaking study described in chapter 4. In that study, it was found that 79 percent of all important innovation designs commercialized had been developed by kayakers and revealed for free. The reduced R&D costs for kayak producers adopting those free designs were very significant: my colleagues and I calculated that development cost savings were 3.2 times larger than whitewater kayak producers' *total* product design budgets over the entire history of that sport (Hienerth, von Hippel, and Jensen 2014).

Producers' support for free innovation

In the previous section we saw that complements and commercializable product designs that spill over to producers "for free" can greatly reduce producers' internal R&D costs. Producers therefore may wish to invest in supporting the design work of free innovators to enhance their supply of free designs. They may do this by supplying free innovators with development platforms and tools that make their design and diffusion tasks easier, and that also guide free innovators' efforts in commercially profitable directions. This is the fourth form of interaction that we see between the free innovation paradigm and the producer innovation paradigm, and is represented by the upward-pointing arrow in figure 6.1.

The empirical literature describes many types of investments by producers to encourage and support innovation by free innovators. Producers may sponsor a user innovation community (West and Lakhani 2008; Bayus 2013) or a design contest (Füller 2010; Boudreau, Lacetera, and Lakhani 2011). They may provide free innovators with kits of tools to enable them to make their own designs more easily (von Hippel and Katz 2002; Franke and Piller 2004). Producers may also engage in boundary-spanning activities and may invest the working time of employees in supporting free innovators (Henkel 2009; Colombo, Piva, and Rossi-Lamastra 2013; Dahlander and Wallin 2006, Schweisfurth and Raasch 2015). Detailed examples of producer support for free innovators, and producer strategy considerations as well, will be discussed in chapter 7.

Modeling Producer Strategies to Support Free Innovation

In line with standard microeconomic modeling, the focus of Gambardella, Raasch, and von Hippel (2016) is on the implications of free innovation *for producers* as well as on the effects of free innovation and producer innovation on social welfare. I will discuss producer innovation strategies in this section and will turn to the implications for social welfare in the next section. In both sections, I will describe the model variables and modeling results conceptually rather than mathematically. The full mathematical model and findings are provided in appendix 2.

Recall from the above descriptions of four paradigm interactions that two of them are positive for the producer. First, producers' profits increase when free innovators create and diffuse complements that producers do not find it profitable to produce and sell, but that enhance the value of the products or services that producers do sell. Second, producers' costs of developing innovations are reduced when they can adopt designs from free innovators instead of developing designs in-house.

Recall too that one of the four paradigm interactions—free-contested markets—is unalloyedly negative for producers: Free peer-to-peer distribution of products or services by free innovators is a source of competition for producers trying to sell the same thing or a substitute. Just like any other form of competition, competition from participants in the free innovation paradigm decreases the size of a producer's market and/or forces the producer to lower prices. For example, in the mountain biking example discussed earlier, free innovator-developed mountain bike designs were available "for free" to mountain bikers—potential customers—just as they were to mountain bike producers. Individuals who elect to build their own bikes reduce the size of the producers' market by removing themselves as potential customers.

Finally, recall that the fourth interaction is the provision of design support by producers to free innovators. This interaction is under producer control, and it is the path by which the model of Gambardella, Raasch, and von Hippel (2016) envisions that a producer can seek to affect and shape the first three interactions to increase profits.

The model's approach to the interplay of the four interactions is to focus on the fraction of a producer's potential market that is capable of both innovation and self-supply. This is because the profitability of a decision to invest in supporting free innovation development turns out to be centrally affected by that factor. (Innovation design and innovation self-supply generally go together. If you are going to go to the trouble of designing something, you will generally build a copy too as part of the development process. If you are a user, the copy you have made will remove you from the producers' potential market—you have supplied it to yourself.)

Suppose that in a particular market very few individuals in the household sector have the capability to innovate in ways that may be of commercial value to a producer. In that case, the model finds, it would make sense for a producer to stick to in-house development and not invest in developing and supplying innovation design tools to support the efforts of just those few free innovators. The cost per additional free innovation developed would be too high. As the fraction of potential free innovators in the producer's market grows, however, investing a portion of producer R&D dollars in tools to support and increase free innovation becomes more profitable than an exclusive focus on in-house development, even if free innovators both innovate and self-supply and thereby remove themselves from that producer's potential market.

Eventually, as the share of potential innovators in a producer's market increases still further, investing in supporting free innovators again becomes unprofitable. The loss of potential market associated with self-supply by potential customers becomes so large that the producer's profits are reduced, even though more free commercializable designs are developed. The offsetting effect of customers' self-production is especially dangerous for producers when *non*-innovating potential customers also gain the ability to make very cheap copies of free innovations. This possibility is a reality today in the case of software and many other information products. Soon, with the increased availability of cheap, personally accessible production technologies such as 3D printers, it also will be a commonplace reality for many physical products.

Of course, this offsetting effect applies only to products that a producer wants to commercialize in competition with peer-to-peer diffusion. In the case of the development of valuable complements that a producer *does not* want to commercialize, the more free innovation and self-supply the better! For this reason, as we will see in the next chapter, today some producers make heavy investments to specifically encourage and support free innovators' development of complements to the commercial products they sell.

I should note that Gambardella, Raasch, and von Hippel (2016) assume that a natural level of free innovation and self-provisioning by potential customers will be present even without a producer firm making investments to support it. As national surveys show, free innovation is a very widespread phenomenon today, generally without intentional support from producers. This implies a possibility, not included in the model, that the natural level of free innovation and self-supply can already be at a point that is "too high" from the point of optimum producers' profits in some markets. Evidence shows that producers judging this to be the case may then choose to invest in frustrating free innovation rather than supporting it. They may, for example, use legal restraints and/or technical barriers to make their products more costly for potential customers to modify or copy (Braun and Herstatt 2008, 2009).

Finally, independent of the number of potential innovators in a market, a producer's best choice with respect to its corporate R&D investments is never to invest only in providing tools to support free innovators. Free designs can seldom be produced commercially just as they are. A producer therefore must invest internal funds to refine a free design and prepare it for production. In addition, a producer must invest internal funds to develop types of designs that free innovators would not be interested in developing but that are important to the market—for example, designs to make products easier for novices to use. The model therefore addresses the appropriate balance with respect to investments that *complement* the efforts of free innovators versus investments that substitute for design development activities that free innovators find it viable to make on their own.

Modeling the Effects of Free Innovation on Social Welfare

Social welfare functions are used in welfare economics to provide a measure of the material welfare of society, with economic variables as inputs. A social welfare function can be designed to express many social goals, ranging from population life expectancies to income distributions. Much of the literature on innovation and social welfare evaluates the effects of economic phenomena and policies on social welfare from the perspective of total income of a society without regard to how that income is distributed. The model presented in Gambardella, Raasch, and von Hippel (2016) takes that viewpoint.

On the face of it, free innovation should increase social welfare. It involves decisions by individuals to divert part of their discretionary unpaid time, generally assumed in economics to be devoted to consumption, to activities that produce value for the innovators themselves, and often produce value for additional peer and commercial adopters too (Henkel and von Hippel 2004).

As markets move from a traditional producer-only situation to a situation including free innovators, the modeling of Gambardella, Raasch, and von Hippel (2016) finds that both producers' profits and social welfare always increase if firms adopt a strategy of investing in complementing free innovation activity instead of competing with it. In contrast, if producers elect to compete with free innovators' designs, both producer profits and social welfare are likely to suffer.

In other words, and as I noted at the start of this chapter, the modeling and theory building my colleagues and I have done concludes that the most profitable and welfare-enhancing situation in the economy involves a division of innovation-related labor between free innovators working within the free innovation paradigm and producers working within the producer paradigm. The optimal division of labor, however, will not be arrived at without policy interventions. As the number of free innovators in markets increases steadily as a result of the technological trends described in chapter 3, our model shows that producers generally switch from a producer-only innovation mode to a mode utilizing free innovation "too late" from the perspective of overall social welfare. The reason is that overall welfare includes benefits that accrue to free innovators and increase social welfare, but that are not taken into account in private producers' calculations of returns.

Producers assess their private returns to investments in supporting free innovation by considering the value they are likely to derive from increased creation of commercially valuable free designs by free innovators. But investments by producers to support free innovation also support the creation of designs that have personal and social value but do not have commercial value. In addition, producers' investments to support free innovation induce other types of self-reward valued by free innovators but not by producers—for example, the learning and enjoyment that free innovators gain from participating in free innovation development. For these reasons, a level of investment supporting free innovation that is *higher than* the level that is optimal for producers' profits always enhances social welfare.

To bring these added sources of welfare into welfare calculations, my colleagues and I argue that calculations of social welfare should include a "tinkering surplus" component. Social welfare is conventionally calculated as profits (PS) plus a consumer surplus (CS). We suggest adding *tinkering surplus* (TS) as a third component to social welfare, consisting of all the net benefits from self-rewards that free innovators gain from developing their innovations. How significant is the omission of the tinkering surplus in conventional welfare calculations? Given the importance of self-rewards to free innovators documented earlier, the omission can be substantial.

Discussion

The most important finding of Gambardella, Raasch, and von Hippel (2016) is that both producers' profits and social welfare are generally increased if producers invest less in capabilities and innovations that *substitute for* what free innovators find it viable to do, and invest more in capabilities and innovations that *complement* free innovation.

For example, in the video game industry, producers should focus their own development efforts on developing game "engines"—a very complex type of software program that has been, at least so far, seldom viable for free innovator developers. In contrast, they should leave the development of simpler and cheaper game "mods" to their gamer customers. Similarly, medical equipment producers may want to leave the pioneering of some new types of medical devices to free innovator patients. (As we will see in chapter 10, free innovating patients are entirely within their legal rights to create, use, and freely share designs for novel medical devices without governmental approvals.) The producers would then focus their R&D investments on the complementary tasks of making the patients' designs better and more reliable through product engineering, and on getting the devices through costly governmental approval processes.

Recall that the modeling also found that, from the perspective of social welfare, producers tend to switch "too late" from a focus on internal R&D only to a division of labor with free innovators as the proportion of free innovators in their markets increases. This is because producers' profit calculations do not take the welfare benefits arising from free innovators' tinkering surplus into account. Novel policy measures may be needed to address this problem. Indeed, some existing policies may make the problem worse and should be reassessed. Policies to subsidize producers to develop innovations that free innovators can also develop will further retard producers' transition to an appropriate division of labor with free innovators. The net effect will be to redistribute welfare from free innovators to firms, and even to lower aggregate welfare.

Again and in summary, my colleagues and I find that both producers and society can benefit from a conscious, intelligently implemented division of innovation labor between innovators acting within the free innovation paradigm and firms acting within the producer innovation paradigm. In the next chapter, I will explore some practical steps in this direction.

7 | Tightening the Loop between Free Innovators and Producers

As the scale of voluntary and unpaid design effort in the household sector becomes clear, both free innovators and commercial project sponsors are increasingly competing to "tighten the loop" between themselves and free innovators to obtain a larger share of that valuable resource. In this chapter, I will first explain how producers are learning to support free innovators in ways that benefit themselves but not their rivals. Next, I will explore how lower-cost pathways to commercialization are becoming available to household sector innovators. Finally, I will discuss how, via crowdsourcing, free innovators and producers are both learning to more effectively recruit free innovation labor from the household sector.

Visualizing the Loop

Recall that there are two pathways involving information and resource transfers between the free innovation paradigm and the producer innovation paradigm. First, innovation designs created by free innovators may be transferred from actors in the free innovation paradigm to actors in the producer innovation paradigm for commercial production and diffusion. Second, producers may transfer tools and other types of support to actors in the free innovation paradigm to assist free innovation development efforts. As we saw in our discussion of modeling findings in chapter 6, these two types of transfer between paradigms are related. That is, innovation support provided by producers can affect the rate and direction of free innovators' efforts, and as a consequence affect the transfer of commercially relevant designs from free innovators to producers. Due to these interactions, we can visualize the arrows between the two paradigms as forming a loop interconnecting activities within each, as is shown in figure 7.1.

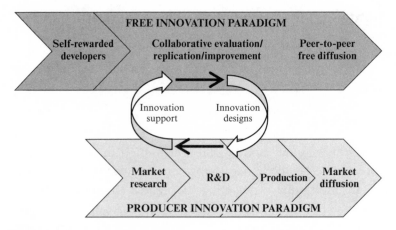

Figure 7.1
Tightening the loop between free innovators and producers (dark horizontal arrows).

Tightening the Loop

In earlier days, before the potential value to producers of free innovation activities was appreciated, any transfer of innovation support from producers to free innovators—one side of the "loop" of interaction between them—was typically accidental. For example, automobile producers might create a car design that was especially easy for customers to modify, and which for that reason attracted a great deal of interest from free innovators. The utility of that design as a platform for free innovator "hacking" probably was not even in the thoughts of engineers working for the producer—at least, in earlier days. They were focused on developing the best designs for large market segments of non-modifying customers.

The second part of the loop—transfer of any designs that were created from free innovators to producers for assessment of commercial potential—was similarly neglected in earlier days, or even actively suppressed. The increased legal risks to auto producers of liability for accidents involving customer-modified products make an effort to suppress understandable (Barnes and Ulin 1984). As a result, even if some innovations generated by free innovators had commercial potential, they were not likely to come quickly or efficiently to the attention of automobile producers' engineering departments.

Similarly, in the earlier days of video games, the possibility of modification of the games by free innovators was not contemplated by video game producers, and the potential commercial value of these was not appreciated. As a result, the games were not designed to be easily modified by gamers, and gamers' innovative activity, if noticed, was discouraged. This reaction was again understandable. Early hacks made by gamers were sometimes designed to parody the commercial game rather than enhance it. For example, *Castle Wolfenstein*, a popular game introduced in 1981, involved fierce combat among dangerous-looking World War II soldiers. Hackers redid it as "Castle Smurfenstein" in 1983, replacing the soldiers with amusingly nonthreatening blue Smurfs (Castle Smurfenstein 2016).

Today, the value of designs generated by free innovators are much clearer to at least some, and producers are responding by "tightening the loop" between the paradigms to increase profits. Indeed, some producers are finding that they can provide design tools and innovation environments to free innovators that do more than promote innovation. They can also shape and channel free innovators' activities toward designs with higher profit potential to specific producers, and also insure that these cannot easily spill over to benefit rivals. As an example, consider a platform set up to support customer innovation by the video game producer Valve (Steam Workshop 2016).

Steam Workshop contains software tools to assist gamers in creating modification to video games. Modifications can involve small changes to games or can be large collaborative efforts that may change a game fundamentally. Certain types of mods, such as creating new game "maps," are specifically supported, thus pulling more free innovator effort into innovation types especially profitable from the perspective of the producer. The total amount of activity by gamers utilizing Steam is quite large. The site claims that more than a million "maps, items, and mods" have been posted on it, and that these have been used by more than 12 million gamers to date (Steam Workshop 2016). Because the postings are on Valve's site, Steam Workshop personnel can monitor the popularity of the various mods posted to gain market insight. Valve can elect to commercialize innovations posted on Steam Workshop and also can elect to financially reward contributors, in that way drawing household sector innovators with producer motives into the mix along with free innovators.

To understand how spillovers to rival producers are avoided on Steam Workshop, consider that video games today consist of application software that "runs on top of" underlying game engine software and is a specific complement to it. The underlying proprietary game engine supplied for the use of free innovators provides such basic video game functions as rendering and animating the objects and characters used in a game. The application software, designed to run on that specific game engine, contains a game's story and setting. Games designed to run on one engine therefore will not run on another—and in this way spillovers are avoided: the free innovations are complements specific to one producer's proprietary game engine (Jeppesen 2004; Henkel, Baldwin, and Shih 2013; Boudreau and Jeppesen 2015).

Other producers profit in similar ways from designs developed by free innovators that do not easily spill over to rival producers. For example, Ikea sells standard modular furniture, each item of which has a specific intended use. Free innovators have learned to modify Ikea furniture in order to use it in ways that the producer did not intend. For example, they might purchase several Ikea picture frames and cut them up to make wall sculptures, or purchase an Ikea bookshelf and modify it to create a fold-out desk. The free innovators then openly share their designs on sites such as Ikeahackers.net. As in the Valve case, the free designs are value-enhancing complements that are specific to Ikea products, and so do not easily spill over to benefit rivals (Kharpal 2014). Again similarly, Lego supports users' creation and sharing of innovative designs made from bricks purchased from Lego. These designs are specific to Lego products, and so do not spill over to benefit rival firms (Antorini, Muñiz, and Askildsen 2012; Hienerth, Lettl, and Keinz 2014).

The Path to Commercialization

Free innovators can also tighten the loop between the free and producer innovation paradigms by electing to become producers themselves (Shah and Tripsas 2007). Two general pathways available to free innovators who wish to become producers are commercializing the design via an existing firm and founding a new venture to commercialize the design.

With respect to commercializing a product via an existing firm, a product "publisher" model of commercialization is emerging to complement the traditional model of product acquisition by firms. For example, when an innovator elects to produce a copy of a product design by using the custom 3D printing service offered by Shapeways, that firm routinely asks the customer if he or she would like to also offer the design for sale to others. The site explains how this works and how it could be attractive: "You design amazing products, we'll help you reach a global market. Start selling today ... Simply design and prototype, and we'll take care of production, distribution, customer service, and all the nitty-gritty. ... No inventory or financial risk. We'll produce and ship your product each time someone orders it, and you keep the profits. ... [We offer] help all along the way. Global Customer Service team, in-depth tutorials, and a supportive community to guide you" (Shapeways 2016).

Founding and funding stand-alone ventures by household sector innovators are also becoming much cheaper and easier than they were in the past. Consider, for example, the fairly recent option to cheaply fund the commercialization of individual products via crowdfunding appeals (Lehner 2013; Mollick 2014). Consider also the steadily improving options for new ventures to outsource costly functions such as product production and delivery to specialized firms.

Baldwin, Hienerth, and von Hippel (2006) describe the typical pathway from free innovation to commercial production. First, one or more household sector innovators create an innovation that turns out to be of general interest. Next, a community grows around that innovation, with each participant self-supplying a copy of the innovation for personal use. Soon, some participants grow to prefer a source of commercial supply for the innovation instead of self-production. As a result, a profitable opportunity for the founding of a new venture arises.

Early responders to such an emerging opportunity for commercialization are generally start-ups formed by some members of the community that grew around an innovation rather than unaffiliated entrepreneurs or pre-existing firms. This is because early information about an innovation and about related commercial opportunities will initially be clearest to participants in such a community. Those individuals are in the best position to know from firsthand participation what is needed and how quickly demand for products responsive to the need is likely to increase. Second, new venture founders from within the community will have an advantage with respect to initial marketing, thanks to their pre-existing relationships with potential customers within their community (Fauchart and Gruber 2011). Of course, existing producers can also seek to gain early insights into emerging commercial opportunities by hiring community members as "embedded lead users" (Schweisfurth and Raasch 2015).

Crowdsourcing

Household sector resources can be tapped directly by both free innovators and producers: producers are not the only ones striving to more effectively tap this resource. Both free innovator and producer project sponsors increasingly seek help from individuals in the household sector through "crowdsourcing." Crowdsourcing is defined as "the act of outsourcing a task to a 'crowd,' rather than to a designated 'agent' ... in the form of an open call" (Afuah and Tucci 2012, 355; Howe 2006). The crowdsourced "task" may range from very general ("Come work on this general topic with us") to very specific ("We need a solution to this specific problem"). Crowdsourcing offers a way to get individuals who are not known to a project's sponsors, but who judge themselves well suited to contribute to solving a specified problem, to identify themselves.

Crowdsourcing calls are attractive to project sponsors for two major reasons. First, it is now understood that calling upon a crowd can sometimes produce better solutions than can calling upon a much smaller set of paid employees to solve a problem. Second, recruiting free household sector labor can often be cheaper than recruiting and paying employees (Agerfalk and Fitzgerald 2008).

The advantages of calling upon a crowd for innovation contributions has been surprising to many, and is at variance with traditional assumptions. It had long been assumed that producer firms would be more effective than unpaid individuals from the household sector in solving innovation-related problems. That assumption was based on the idea that larger-scale R&D organizations can afford to hire very specialized and expert developers, and also can economically justify expensive specialized R&D equipment to increase the problem-solving efficiency of those employees still further.

However, it is now more deeply appreciated that the better problem-solving performance of experts can be quite narrow (Larkin, McDermott, Simon, and Simon 1980; Gobet and Simon 1998). An expert developer of jet engines, for example, may be no better than a novice at designing other types of propulsion devices. Therefore, especially in the case of development problems for which one does not already know the type of solution one is seeking, asking the crowd can offer a very important advantage. Although the expertise of individual developers in the crowd may be just as specialized as that of individual employees of the producer, the crowd collectively will have a very wide range of expertise to call upon via crowdsourcing. In line with that supposition, it has been shown that opening access to a problem to a wide range of individuals having highly diverse information via a crowdsourcing call can contribute greatly to solving some problems in creative ways (Raymond 1999; Benkler 2002, 2006; Frey, Lüthje, and Haag 2011; Jeppesen and Lakhani 2010). An additional advantage is that information about pre-existing *solutions* may also exist within the crowd. In fact, information on pre-existing solutions suited to a new problem may make up much or most of the useful information that crowdsourcing provides. Lakhani, Jeppesen, Lohse, and Panetta (2007), in a study of winning solutions in crowdsourcing contests sponsored by the firm Innocentive, found that 72.5 percent of winning solvers' submissions were based partially or entirely on previously developed solutions. Pre-existing solutions, being better understood, can be preferable to entirely new solutions.

With respect to the second point, household sector contributors to producer innovation projects can be cheaper as well as better performing than firm employees because, as has been discussed, free innovators are largely self-rewarded. Some research into why consumers are willing to participate in crowdsourced innovation activities without monetary compensation has been done, and more is being done. (See, e.g., Nambisan and Baron 2009; Kohler, Füller, Matzler, and Stieger 2011; Yee

2006; Stock, von Hippel, and Gillert 2016.) As the nature of self-rewards desired by potential contributors becomes better understood, project sponsors will be able to more efficiently and effectively provide exactly those rewards. Conditions under which individuals in the household sector are willing to participate "for free" in a project profitable for producers are also being studied. For example, it has been found that a system offering clear benefits to producers must be seen as "fair" by potential contributors if it is to be effective and sustainable (Franke, Keinz, and Klausberger 2013; Faullant, Füller, and Hutter 2013; Di Gangi and Wasko 2009). All this ongoing work will enable steady improvements to crowdsourcing practices.

Three examples of crowdsourced projects, one sponsored by free innovators, one sponsored by scientists, and one sponsored by a producer, will illustrate the broad applicability of crowdsourcing within and beyond free innovation projects.

Nightscout, a free innovation project

An example of a crowdsourcing call by free innovators is the Nightscout project described in chapter 1. Recall that this free innovation project is devoted to the development and distribution of improvements to medical devices used by diabetes patients. Note the implicit call for additional volunteer effort in the project description text posted on the Nightscout webpage:

Nightscout was developed by parents of children with Type 1 Diabetes and has continued to be developed, maintained, and supported by volunteers. When first implemented, Nightscout was a solution specifically for remote monitoring of Dexcom G4 CGM data. Today, there are Nightscout solutions available for Dexcom G4, Dexcom Share with Android, Dexcom Share with iOS, and Medtronic. The goal of the project is to allow remote monitoring of a T1D's [Type 1 diabetic's] glucose level using existing monitoring devices. (Nightscout project 2016.)

Foldit, a citizen science project

As an example of a crowdsourcing call for free household sector contributions to a citizen science project, consider Foldit. Foldit is a project developed and sponsored by scientists from the University of Washington to study how proteins fold in nature. Needing many specific protein-folding solutions as inputs to their research, the scientists sought free help from "the crowd." Because people in the household sector do not have a personal use for protein-folding solutions, the scientists sought to attract participants by offering other forms of self-reward. Specifically, they designed their project to offer the self-rewards common to games played for pleasure, utilizing gamification design practices (Zicherman and Cunningham 2011):

To attract the widest possible audience for the game and encourage prolonged engagement, we designed the game so that the supported motivations and the reward structure are diverse, including short-term rewards (game score), long-term rewards (player status and rank), social praise (chats and forums), the ability to work individually or in a team, and the connection between the game and scientific outcomes. (Cooper, Khatib, Treuille, Barbero, Lee, Beenen, Leaver-Fay, Baker, Popovic, and Foldit players 2010, 760.)

The Foldit game is difficult, requiring online training sessions before productive play can begin. Still, the scientists were successful in attracting many people to help, with 46,000 volunteers playing Foldit during their unpaid, discretionary time in 2011. The work these volunteers contributed was very valuable to the project's sponsors, providing specific protein-folding solutions and also providing new methodological insights that were then used to improve computerized folding algorithms.

The scientist-developers of Foldit conducted a small, informal survey asking contributors why they had chosen to participate in Foldit (Cooper et al. 2010). Forty-eight players responded with up to three reasons each. As would be expected in view of the subject matter, use and sale motives were entirely absent. About 30 percent of respondents reported that immersion (e.g., "it is fun and relaxing") was important; 20 percent mentioned achievement (e.g., "to get a higher score than the next player"); 10 percent mentioned social benefits (e.g., "great camaraderie"); 40 percent reported being motivated by a wish to support the purpose of the project (e.g., [I wanted to help] "to crack the protein folding code for science") (supplement to Cooper et al. 2010, 12). These self-rewarding motives probably are similar to those involved in other forms of charitable giving: One gives in part "to help others" and in part to support a specific cause of high personal interest (Webb, Green, and Brashear 2000).

A producer crowdsourcing project

Swarovski, a jewelry producer, wanted to attract consumers to devote discretionary time to designing novel and fashionable jewelry. With the help of Hyve, a company that specializes in building online problem-solving sites, Swarovski created a crowdsourcing site that offered volunteer participants the opportunity to develop their own jewelry designs, to showcase them, to comment on and vote on the designs of others, to upload their avatars and photos, and to be included as a trendsetter in a book about trends in watch design (Füller, Hutter, and Faullant 2011). Participants had no expectation of seeing their designs produced and no expectation of payments related to commercialization of their designs. Nonetheless, the initiative to attract participation from the household sector was successful. More than 3,000 designs were uploaded by more than 1,700 participants.

Füller (2010) surveyed contributors to ten different virtual co-creation projects hosted by Hyve, the subjects including designing a baby carriage, furniture, mobile phones, backpacks, and jewelry. He found that the motivators of "intrinsic innovation interest" and curiosity were most important to survey respondents: "In contrast to open source communities and user innovations, where members engage in innovation tasks because they can benefit from using their innovation, consumers engage in [Hyve] virtual new product developments mainly because they consider the engagement as a rewarding experience" (Füller 2010, 99).

Discussion

Today, sponsors of both free innovation projects and producer innovation projects are competing increasingly strongly for the discretionary time and resources of individuals in the household sector. And, as we have seen, producers are learning to more skillfully "tighten the loop" that can profitably connect the free and producer innovation paradigms. How this competition will play out will only be seen over time.

Innovation projects sponsored by producers may become systematically more appealing to many household sector contributors than free innovation projects. Producers, after all, may be willing to invest more than free innovators to understand and enhance self-rewards desired by individuals in the household sector. This might, in turn, reduce free effort available for innovation pioneering. For example, individuals attracted to Valve's skillfully gamified Steam Workshop, and encouraged by the tools offered to create yet another "mod" for an existing video game, may be drawn away from developing fundamentally new forms of digital entertainment.

Alternatively, it may be that some free innovators cannot be attracted to producer-supplied tools and platforms, and will instead elect to develop and use free tools. We see this pattern illustrated today in the case of the development of new statistical tests and methods. There are well-known commercial statistical software packages, like SPSS and Stata, that are purchased and used by many. The producers of these include toolkits in their products to enable their customers to develop new statistical tests within the commercial program—much as Valve offers game mod development tools to its customers. However, many innovative statisticians find these toolkits, shot through with producer constraints intended to protect proprietary advantage, to be unacceptably constraining. These individuals therefore often opt to do their development work on a free, open source statistical software platform named R (r-project.org). Here, they have full creative freedom to study and modify the core program, and also to develop and freely share new tools and new statistical tests with peer developers. This pattern frees free innovators from producer constraints. At the same time, it need not greatly disadvantage the commercial producers. Although producers cannot exclude rivals as they may be able to do in the case of tests developed using commercial toolkits, they can still obtain the advanced tests developed within R for free and, with some adaptations, incorporate them into their commercial products.

In the end, producers may find that offering less constraining toolkits to free innovators has commercial advantages. Thus, it has been found that producers that more broadly empower consumers to innovate are rewarded by stronger marketplace demand for their products (Fuchs, Prandelli, and Schreier 2010; Fuchs and Schreier 2011).

To date, empirical studies of free user innovation have focused almost entirely on product innovations. However, free innovation logically should extend far beyond products. After all, the test for innovation opportunity viability presented in chapter 3 has nothing to say about the nature of specific opportunities. It just specifies that innovators' expected benefits should exceed their expected costs.

In this chapter, I show that the scope of free innovation in the household sector is indeed broad—and perhaps as broad as that of producer innovation with respect to products, services, and processes of interest to consumers. I do this by reviewing the findings of field-specific studies and by discussing illustrative examples of the sources of innovation across five innovation categories used in official OECD government statistics.

Types of Innovations

To test for the ubiquity of free innovation, I use the definition of innovation used by government statistical agencies in OECD nations. "*An innovation is the implementation of a new or significantly improved product (good or service), or process, a new marketing method, or a new organizational method in business practices, workplace organization or external relations*" (*Oslo Manual* 2005, paragraph 146, italics in original). Adjusting that producer-centric language to include the possibility of free innovators, we see that it refers to five innovation subject matters: An innovation is a new or significantly improved (1) product, (2) service, (3) process, (4) marketing method, or (5) organizational method related to free or producer innovation practices or external relations.

In the sections that follow, I briefly document the presence of free innovation in the household sector with respect to Oslo Manual categories 2-5. The importance of free innovation activity in category (1), products, has already been documented in earlier sections of this book.

Free user innovation in services

Uniform governmental statistics on services are collected under nine high-level categories: wholesale and retail trade; hotels and restaurants; transport, storage, and communication; financial intermediation; real estate, renting, and business activities; public administration and defense; education; health and social work; and other community, social, and personal service activities (UN 2002). Services are of great economic significance. Taken together, all services make up a portion of GDP that is roughly twice as large as that of all products.

There are two main attributes that distinguish services from products. In the case of a service, (1) production and consumption cannot be separated and therefore (2) one cannot keep a service in inventory (Fitzsimmons and Fitzsimmons 2001; Zeithaml and Bitner 2003; Vargo and Lusch 2004; Crespi, Criscuolo, Haskel, and Hawkes 2006). In contrast, one can do both of these things in the case of a product. For example, a producer can build a taxi and put it into inventory to await a buyer. A taxi is a product, and production and consumption of products can be separated. However, a taxi *ride* is a service, and so a provider cannot similarly offer, available for purchase, an inventory of completed rides from your workplace to your home. Alas, one must patiently sit in the cab, consuming a ride exactly when and as it is produced. The same is true of medical services. Again, and again alas, one cannot purchase a completed medical operation; one must consume it as it is produced.

Services are often thought of as necessarily involving a provider and a consumer (Vargo and Lusch 2004). For example, a taxi service involves both a driver and a passenger, the passenger receiving the transportation service and taxi driver (or self-driving taxi) providing it. But it is also true that a passenger can drive himself or herself—that is, self-provide a similar transportation service. When consumers *can* "serve themselves," it is also possible for them to innovate with respect to the services they deliver to themselves. Just as in the case of products, these services may then diffuse to peers as DIY self-services, and also may diffuse to producers for commercialization.

In subsections that follow, I will summarize the findings of three empirical studies of the sources of innovation in three types of services: retail banking, mobile banking, and medicine. As will be seen, service innovation development by free innovators is prominent in all three.

Free user development of retail banking services Oliveira and I studied the sources of commercially important services in retail banking (Oliveira and von Hippel 2011). The sample consisted of all basic types of retail banking services offered by major banks in 2011 within the traditional range of "core" banking services, such as loans, checking accounts, savings accounts, and time deposits. Services offered beyond that range, such as brokerage and insurance services, were excluded. Within the core banking services, we focused on innovations that had been first commercialized by retail banks between 1975 and 2010.

During the period Oliveira and I studied, banks were introducing new services in computerized rather than manual forms. For many of the sixteen major retail banking services first commercialized between 1975 and 2010, however, there had been earlier manual ways to perform the essentially the same services. To understand the full innovation history, we therefore sought to identify both the developers of the first computerized version of each service in our sample, and also the developers of the manual precursors to those services where that information could be found.

Table 8.1
Sources of important retail banking services.

Service type	n	free user	bank	joint
Developers of manual precursor services	10	80%	0%	20%
Developers of first computerized versions	16	44%	56%	0%

Source: Oliveira and von Hippel 2011, tables 3 and 4

As can be seen in table 8.1, my colleague and I determined that 80 percent of the manual precursors for the basic types of computerized services offered by major banks today were developed by household sector users who had personal uses for those innovations. User innovators also developed of 44 percent of the first computerized versions of those services. As best we could tell from searches of the literature and from interviews of experts, all were free innovations, not protected by intellectual property rights, and available for free adoption. As an example of a basic service for which both the manual practice and the first computerized version were developed through free innovation by users, consider "account information aggregation." The need for that service arises because many retail banking customers deal with multiple banks or other financial institutions at the same time. For example, your checking and savings accounts might be with one bank, your home mortgage may be serviced by another, and your credit card accounts may be serviced by still other banks. Somehow, financial information from all these institutions must be "aggregated" so that you can see and manage your overall financial situation.

Until 1999 each bank reported to each customer only its own financial dealings with that customer. Customers then aggregated multiple reports from multiple banks for themselves, using their own methods, and so were the initial developers and users of manual versions of "account information aggregation." Individuals also were the first to develop the computerized version of this service in the basic form that was eventually commercialized by banks. Consider this individual's personal innovation history:

I do my banking online, but I quickly get bored with having to go to my bank's site, log in, navigate around to my accounts, and check the balance on each of them. One quick Perl module (Finance::Bank::HSBC) later, I can loop through each of my accounts and print their balances, all from a shell prompt. With some more code, I can do something the bank's site doesn't ordinarily let me do: I can treat my accounts as a whole instead of as individual accounts, and find out how much money I have, could possibly spend and owe, all in total. Another step forward would be to schedule a cron entry (Hack#90) every day to use the HSBC option to download a copy of my transactions in Quicken's QIF format, and use Simon Cozens' *Finance::QIF* module to interpret the file and run those transaction against a budget, letting me know whether I'm spending too much lately. This takes a simple web-based system from being merely useful to being automated and bespoke; if you can think of how to write the code, you can do it. (Hemenway and Calishain 2004, 62)

The computerized information aggregation service now offered commercially by banks functions in essentially the same way as this individual's version. With an account owner's permission, a bank automatically contacts each financial institution with which a retail user has an account, logs on with the user's password, collects information on the status of each account, and logs off. It then assembles the information collected from all accounts into a spreadsheet tailored to the user's specifications.

Free user development of mobile banking services   Mobile phone banking is based upon a technically very sophisticated cell phone platform. Despite this, the platform offers novel service possibilities that can be discovered by individuals who do not understand its technical details. (By analogy, innovators can and do develop important new uses for airplanes, e.g., carrying the mail or spotting forest fires, without having to know in any technical detail how an airplane actually functions.) Van der Boor, Oliveira, and Veloso (2014) examined the histories of a complete list of the twenty basic mobile financial services reported by Groupe Speciale Mobile Association (GSMA). They found that 85 percent of these innovations originated in countries with relatively poor conventional retail banking service infrastructures, where the need was high. They also found that 45 percent were first developed by household sector users. Cell phone service providers developed 45 percent, and 5 percent were developed by users and producers jointly. One (5 percent) was developed by a firm with a business use for the innovation.

As a typical innovation history, consider the development of a method for transferring money—a basic mobile banking service—by cell phone users in the Philippines. In the Philippines, customers could pay for their cell phone use by means of "scratch cards" sold at retail stores. After buying a scratch card of a certain denomination, the purchaser was instructed to scratch an obscuring layer from the surface of the card to reveal a unique multi-digit activation code. When typed into the phone, that code transferred prepaid cell phone credit to that customer's phone number.

In 1998, customers in the Philippines recognized that they could also use scratch card codes for a fundamentally different purpose.

Instead of adding minutes ("airtime") to their own phones, they could transfer the credit codes to others as an acceptable substitute for cash. To accomplish this, the purchaser of a scratch card, instead of entering the activation code revealed on the card into his own phone, would send the unique activation code by text messaging to a person to whom he wished to transfer money. That person could then use the paid-for airtime, or pass the credit along further as he or she chose. As a second basic service, individual users subsequently pioneered the use of airtime as a form of currency for merchant payments. Five years later, in 2003, cell phone service producers began to offer commercial versions of these banking services, which by then were already in widespread consumer use (van der Boor, Oliveira, and Veloso 2014). All of these user-developed novel services were unprotected and freely shared, and thus meet the criteria for free innovations.

Free user development of medical services for patients with rare diseases There are between 5,000 and 8,000 rare diseases that, taken together, afflict approximately 8 percent of the world's population (Rodwell and Aymé 2014; Committee for Orphan Medicinal Products and European Medicines Agency Scientific Secretariat 2011). Many of these diseases are chronic and impose significant difficulties on the daily lives of both patients and their caregivers (Song, Gao, Inagaki, Kukudo, and Tang 2012). Small market size, due to the low prevalence of each disease, makes it commercially unattractive for pharmaceutical firms and other medical suppliers to invest in development of new products and services specifically for a rare disease (Acemoglu and Linn 2004). As a consequence, patients with rare diseases tend to be underserved both clinically and commercially (Griggs, Batshaw, Dunkle, Gopal-Srivastava, Kaye, Krischer, Nguyen, Paulus, and Merkel 2009).

Because patients with rare diseases are often underserved, colleagues and I speculated that they would often decide to innovate to help themselves. To explore this idea, we conducted a survey on that topic among 500 afflicted medical patients in Portugal, using a questionnaire quite similar to the one used in the national surveys described in chapter 2. We found that there was a great deal of self-help innovation among patients with rare diseases and their non-professional caregivers. Of 500 respondents, 36 percent reported developing something they viewed as novel. They also reported on average that their innovations significantly aided them in dealing with their disease and improving their quality of life. Almost all of the innovations were medical services rather than devices. After application of the novelty screening criteria used in the national surveys of product innovation described in chapter 2, 8 percent of respondents (40 of 500 respondents) were found to have developed innovations that were judged by expert medical evaluators to be new to medical practice (Oliveira, Zejnilovic, Canhão, and von Hippel 2015).

As an illustration of a patient-developed service innovation novel to medicine, consider a development by Joaquina Teixeira, the mother of a child with Angelman's Syndrome, a rare genetic disorder. One attribute of Angelman's Syndrome is ataxia, an inability to walk, move, or balance well. Young children with that disability understandably do not want to practice standing and walking and, unless energetic interventions are applied, will not do so. Professional medical advice to parents is simply to "make your child stand and walk often." In practice, following this advice leads to many unhappy interactions between determined parents and reluctant children.

Joaquina Teixeira, who was struggling with exactly that problem, noticed that her son, when at a neighbor child's birthday party, kept reaching for colorful helium balloons that were floating in the party room, high above his head and out of reach. She promptly went and bought 100 helium balloons and released them in a room in her own home. As he had done at the party, her son kept reaching for the strings of the balloons. Teixeira carefully set these strings to a length that he could reach only by standing. He was thus motivated to repeatedly attempt to stand without prompting. His mother constantly varied the challenges, the child never tired of the game, and his standing and walking skills improved greatly. This medical service innovation is easily replicable and was freely revealed by the developer, both in person and via the Internet, to assist other parents and children in the same situation (Teixeira 2014).

Free user development of process equipment: 3D printers
Like commercial producers, free innovators use production processes to create personal copies of the innovations they develop. These

production processes must be quite inexpensive to be within the personal means of individual innovators in the household sector. Production equipment made for producers is often quite costly, and so it is reasonable that free user innovators would attempt to develop less costly production process innovations and improvements for themselves.

Consider the development of personal 3D printers—fabrication machines that use design information encoded in software to "print" physical objects. The major role of free user innovators in the innovation history of this field has been reported by de Bruijn (2010) and by de Jong and de Bruijn (2013).

The innovation history of the 3D printer field (often called additive manufacturing) began in 1981 when Hideo Kodama of Nagoya Municipal Industrial Research Institute invented fabrication methods that built up a three-dimensional object from successive layers of a polymer hardened by exposure to ultraviolet light. Other researchers followed, developing alternative methods of "3D printing," and in 1984 commercial production of 3D printers began. The first commercial machines were quite expensive, costing about $250,000 each. They were marketed to research institutions and to the R&D departments of firms, and were used for rapid fabrication of product prototypes. The time savings over conventional prototype fabrication techniques made the machines quite cost-effective for producers in that application.

In 2004, Adrian Bowyer, a senior lecturer in mechanical engineering at the University of Bath, proposed the development of a rapid prototyping machine that he called RepRap (meaning Replicating Rapid prototyper). Bowyer wanted to design a 3D printer that would be very simple, cheap, and at least partially self-replicating (in the sense that one printer could print many of the parts needed to make additional printers). After his initial proposal, development commenced at the University. The evolving design was openly shared online and soon captured the interest of a widely distributed audience of free innovators who joined the design effort and pooled their contributions. Fewer than ten people were involved in the first year, but interest grew rapidly. By October of 2010, the online hobbyist 3D printer community had grown to between 4,000 and 5,000 participants (de Bruijn 2010, 19, 31).

De Bruijn surveyed 376 members of this online community to determine, among other things, how much time members were spending on various activities related to their hobby. On average, he found, they were spending 10.41 hours working with or developing their personal 3D printing machines per week. That time was distributed into the several activity categories shown in table 8.2. As can be seen, developing improvements to the personal 3D printer—either to print what an individual user wanted or just to make the machine better—accounted for 15 percent of the time devoted by household sector users to activities related to 3D printers. Many important improvements resulted, and all were shared openly. The developers in the online community were free innovators intentionally following open source software community practices (de Jong and de Bruijn 2013).

Table 8.2
Time, per week, spent by the average individual on using and improving a personal 3D printer.

	Hours	Percentage of time
Building the machine	4.9	47%
Printing objects	1.7	16%
Developing improvements	1.5	15%
Helping other users	0.9	9%
Improving skills	1.4	13%
Total	10.4	100%

Source: de Bruijn 2010, table 4.3

Free user development of a "marketing method": community brands
Although free innovators give their innovations away rather than sell them, they can still be interested in marketing methods for a number of reasons. Innovation communities may, for example, wish to advertise for contributors to join their efforts. According to Dahlander (2007, 930), "at times of stiff competition between communities, attracting a base of users and developers is not easy." In addition, they may wish to increase the diffusion of their innovations, motivated by one or more of the various forms of self-reward I discussed in chapter 5.

One example of innovation in marketing methods by free innovators is the use of everyday activities to generate powerful brands at no incremental cost. A brand is a "name, term, sign, symbol, or design, or a combination of them intended to identify the goods and services of one seller or group of sellers and to differentiate them from those of competition" (Kotler 1997, 443). In legal terms, a brand is a trademark. Brands and marketing methods are typically associated with sellers, as in Kotler's definition. However, it is clear that the functions of a brand with respect to identifying the developer of an innovation and that innovators' reputation for quality would be useful to potential adopters of free innovations as well.

Studies show that open source software development communities generate their own powerful brands at no cost by simply creating and displaying a logo or a trademark with which people associate positive experiences both within and outside the community. How does this work? Consider that the general mechanism behind the strengthening and the shaping of brands involves linking similar positive associations to brand names or symbols within the minds of many potential customers (Edwards 1990; Zajonc 1968; Keller 1993). If the effort required to embed mental associations in the minds of many is undertaken for that special purpose and is expensive as in the case of many producers' marketing campaigns— it is not cheap to hire a famous athlete to pose at the top of a mountain holding a branded can of soda—the creation of a brand will be expensive. If, however, the stimulus for a broadly shared mental association arises as a side effect of activities or experiences undertaken for other purposes, brand creation can be costless.

Collaborative free innovation projects often adopt names and logos to demark their projects (for example, the Apache feather, the Linux penguin). As a consequence, community contributors will have the shared experience of working on innovations and interacting with like-minded others with a clear association to the community's logo and name. In the course of their activities, they gain rich positive experiences that are associated with the community and that contain elements similar to those experienced by other community members. The resulting shared mental associations, gained as a byproduct of common activities, should function to costlessly create and strengthen a brand.

In work reported by Füller, Schroll, and von Hippel (2013), two colleagues and I tested this idea by conducting an empirical study of brand strength of Apache and Microsoft Web server software. We found that Apache was the stronger brand both within the Apache community and outside it with respect to that type of software. Interviews with Apache Foundation leaders documented that there was no investment made by Apache to specifically to create or strengthen the Apache brand. This is not a single-case phenomenon, nor is it restricted to open source software. Pitt, Watson, Berthon, Wynn, and Zinkhan (2006) note that the open source movement has produced a series of well-known brands, including Linux and Mozilla Firefox. More generally, Cova and White (2010) term communities that create their own brands "alter-brand" communities.

New organizational methods

Finally, we come to the *Oslo Manual*'s inclusion within official innovation statistics of *"a new organizational method in business practices, workplace organization or external relations"* (*Oslo Manual* 2005, paragraph 146). Individuals acting within the free innovation paradigm have developed many novel ways to work together as unpaid innovators and to collaborate in developing and diffusing innovations. I am not aware of any systematic studies of this particular category of free user innovation, but there are many examples. Participants in open source software projects have been especially active in developing new methods of working together (von Krogh, Spaeth, and Lakhani 2003; O'Mahony and Ferraro 2007; O'Mahony 2007).

As one important example, consider the General Public License, invented by Richard Stallman (2002). In 1985 Stallman, a brilliant programmer in MIT's Artificial Intelligence Laboratory, set about developing and diffusing a legal mechanism that could preserve free access to the software developed by software "hackers." Stallman's innovative idea was to use the existing mechanism of copyright law to that end. Software authors interested in preserving the status of their software as "free" could use their own copyright to grant licenses on terms that would guarantee a number of rights to all future users and innovators. They could do this by simply affixing to their software a standard license that conveyed those rights.

The basic license that Stallman developed to implement that idea was the General Public License (GPL), sometimes referred to as "copyleft." Basic rights transferred to those possessing a copy of free software include the right to use it at no cost, the right to study its source code, the right to modify it, and the right to distribute modified or unmodified versions to others at no cost. Licenses conveying similar rights were developed by others, and a number of such licenses are currently in use. The GPL is a fundamental "organizational method" innovation, developed for the free and open source software field but containing novel ideas and principles that are widely applicable (Torrance 2010; Torrance and Kahl 2014).

As a second example of an important organizational innovation developed by free innovators, consider distributed revision control packages, which are widely used in software development today. Initially created by open source software project developers to manage their own projects, the currently most popular version of such software is GIT, which was initially developed by Linus Torvalds for Linux kernel development in 2005, and which has since been further developed by many others. GIT has spread to many other open source software projects and to many other types of applications via hosting services such as GitHub.com (Ram 2013). GIT makes it possible for all contributors to collaborative efforts to work asynchronously and to merge their contributions at any time. Tools commonly available within GIT and other software packages support the tracing of errors and the maintenance of a full audit trial of past versions. Version control software is an important organizational innovation. Developed by free innovators for their own use within collaborative projects, the principles are widely applicable beyond the organization and management of open source software projects.

Discussion

At the start of this chapter, I argued that innovation opportunities viable for user innovators are likely to extend to many types of innovation in addition to product development. That seemed reasonable because there is nothing in the definition of a viable innovation opportunity that restricts free innovators to product innovation or any other specific kind of innovation (Baldwin and von Hippel 2011). And indeed, we now see that free innovation in the household sector is present within all five basic innovation categories currently measured in OECD nations' innovation statistics (*Oslo Manual* 2005, paragraph 146).

I conclude, from these early empirical findings, that free innovation is likely to be an important contributor to innovative advances across the entire spectrum of innovation opportunities of interest to individuals in the household sector. This is a very valuable result with respect to improving our understanding of the importance of free innovation, and with respect to learning to both measure and utilize it more effectively.

9 | Personality Traits of Successful Free Innovators

Recall from the national surveys summarized in chapter 2 that from 1.5 percent to 6.1 percent of individuals in six countries develop new or modified products for their own use. This is in some ways an impressive figure, representing tens of millions of free innovators in just those six countries. But another way to look at it is that 94–98 percent of individuals in those countries are *not* free innovators, or perhaps try to innovate but fail. Two questions then arise: Are there differences between individuals who successfully carry out innovation projects in the household sector and those who do not? And, if there are differences, can we do anything to increase the amount of successful free innovation?

In this chapter, I draw upon a study by Stock, von Hippel, and Gillert (2016) to identify personality traits significantly associated with successful free innovation in the household sector. Based on these findings, my colleagues and I suggest two possible ways to increase the amount of successful free innovation.

Design of the Study

Given the documented importance of free innovation, it clearly will be valuable to learn more about the characteristics of free innovators. Stock, von Hippel, and Gillert (2016) began this work by a conducting a study of free user innovators' personality traits that are related to innovation success among a sample of 546 German consumers. Our study focused on three successive innovation process stages: (1) having an idea for an innovation for personal use; (2) building a prototype for personal use; and (3) diffusing the innovation either by free, peer-to-peer transfer or to a producer firm. To be able to compare success and failure at each stage, we grouped participants according to how far each had progressed in the innovation process. As can be seen in figure 9.1, progressively fewer consumers successfully completed each successive stage. This allowed my colleagues and me to conduct a "success-failure" comparison at each stage. That is, starting at the left in figure 9.1, we were able to compare the personality traits of those not having an idea (stage 0) with those who did have a product innovation idea (stage 1). Next, we could compare the personality traits of those who did not prototype their idea with those who succeeded in creating a prototype for personal use (stage 2). Finally, we could compare the personality traits of those who did not diffuse their prototyped innovation to the traits of those who successfully did so (stage 3).

The design of our study approximates the real-world situation faced by individual household sector innovators (consumers) in an interesting way. Personality traits are stable, and so those traits an individual has that are associated with success in early stages are necessarily carried into later stages, where those same traits may be less helpful or may even be a hindrance. Conversely, if a trait that enhances individuals' chances of success at, say, stage 3 is negatively associated with success at phase 1, those possessing that trait are unlikely to reach stage 3.

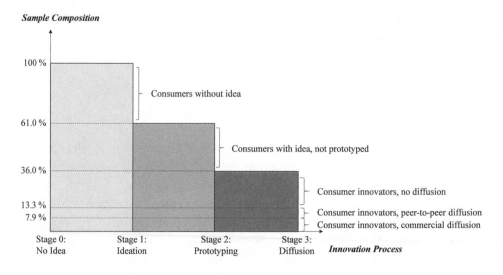

Figure 9.1
Data analysis strategy: comparison of individuals accomplishing vs. not accomplishing successive stages in the innovation development and diffusion process.
Source: Stock, von Hippel, and Gillert 2016, figure 1.

Study Methods

As was mentioned above, the study sample of Stock, von Hippel, and Gillert consisted of German householders. To ensure that we would have enough subjects for analysis at in all three innovation process stages, my colleagues and I recruited participants in two different ways. First, we used "snowball sampling" (Goodman 1961). In this method, individuals who have a rare characteristic—in our study, engagement in innovation development—are asked to identify others they may know who have the same characteristic (Welch 1975). (The utility of snowballing stems from the observation that people with rare characteristics tend to know or be aware of people similar to themselves.) In the second method, our goal was to increase the number of individuals in the sample who were likely to have successfully completed all three stages in the innovation process. We therefore deliberately sought out individuals who either had posted a description of an innovation they had developed on the Internet or had been featured on a German TV program devoted to individual inventors.

The net result was a sample containing both free innovators and entrepreneurial household sector innovators hoping to commercialize their innovations. In total, the sample we recruited for the study consisted of 546 individuals in the German household sector, 443 of them identified via the first method and 103 via the second. The two subsamples of respondents were similar in their demographic characteristics and were combined for analysis. Data were collected from respondents by means of an online questionnaire.

Personality Traits

Personality "traits" are aspects of individuals' personalities known to be highly stable over time, situations, and social roles. Today, studies of personality traits typically use what is called the five-factor model of personality (also known as the Big Five model) consisting of five underlying traits in personality that display minimal overlap. The Big Five model's variables have been proven to encapsulate many more detailed personality variables well, and to be quite stable (Costa and McCrae

1988, 1992, 1995; Goldberg 1993; McCrae and John 1992; McCrae and Costa 1997).

Big Five analyses describe individuals' personalities by the extent to which they display each of five traits in their lives (Barrick and Mount 1991):

• Openness to experience "characterizes someone who is intellectually curious and tends to seek new experiences and explore novel ideas" (Zhao and Seibert 2006, 261; Barrick and Mount 1991). Individuals high on the trait of openness can be described as creative, innovative, imaginative, reflective, and untraditional. In contrast, individuals low on openness prefer the plain, straightforward, and obvious over the complex, ambiguous, and subtle (McCrae and Costa 1987).

• Extraversion "describes the extent to which people are assertive, dominant, energetic, active, talkative, and enthusiastic" (Zhao and Seibert 2006, 260; LePine and Van Dyne 2001; Lucas, Diener, Grob, Suh, and Shao 2000). Those with low levels of extraversion (i.e., introverts) prefer nonsocial situations and are quieter, more reserved, and more independent than those with higher levels (Zhao and Seibert 2006, 260).

• Conscientiousness "indicates an individual's degree of organization, persistence, hard work, and motivation in the pursuit of goal accomplishment" (Zhao and Seibert 2006, 261). Individuals with high scores on conscientiousness have a preference for planned rather than spontaneous behavior (Barrick, Mount, and Judge 2001).

• Agreeableness describes an individual's interpersonal orientation. Agreeable individuals are modest, trusting, forgiving, altruistic, and caring. They tend to conform to social conventions and to engage in cooperative, high-quality interpersonal interactions (Barrick and Mount 1991; Zhao and Seibert 2006). Someone at the very low end of the dimension can be characterized as self-centered, suspicious, and hostile (Feist 1998).

• Neuroticism "represents the tendency to exhibit poor emotional adjustment and experience negative affects, such as anxiety, insecurity, and hostility" (Judge, Bono, Ilies, and Gerhardt 2002, 767; also see LePine and Van Dyne 2001). The opposite of neuroticism is emotional stability.

Study Findings

All the findings of our study are summarized in table 9.1. In the top half of the table, the significances of four "control variables" are presented. In the bottom half, significant relationships between Big Five personality factors and likelihood of success at each innovation process stage are shown.

Findings regarding control variables

In order to see the effects of personality traits clearly, one has to "control for" the effects of other variables known to have a strong relationship to innovation process success; hence the term control variables. (By including control variables explicitly in our study model, we addressed what is called omitted-variable bias. This would result from the absence of an independent variable correlated with both the dependent variable and one or more included independent variables.)

The effects of the first two control variables in table 9.1 have been studied and found important in the national surveys of consumer innovation described in chapter 2 (von Hippel, Ogawa, and de Jong 2011; de Jong 2013; de Jong, von Hippel, Gault, Kuusisto, and Raasch 2015; and Kim 2015). In line with the findings of those studies, on the first row of the table we see that male gender is significantly associated with both successful idea generation and prototyping. Gender may also be statistically associated with successful diffusion. However, because most of the individuals who had succeeded at the earlier phases and so were entering the third and final diffusion stage were male, there was not enough variation in the sample entering stage 3 to assess the significance of that control variable in the diffusion stage.

In the second row of table 9.1, we next see that technical background is significantly associated with successful idea generation. As was the case with gender, technical background is so strongly associated with successfully passing stage 1 that most of the individuals who move on to further stages have technical backgrounds. For this reason, the importance of technical background to success at stages 2 and 3 cannot be analyzed. However, we do know from other research that technical background is also very important to stage 2 prototype development (Lüthje, Herstatt, and von Hippel 2005).

Table 9.1

Effect of personality on the likelihood to successfully pass through stages of the household sector innovation and diffusion process.

	Ideation (stage 1)	Prototyping (stage 2)	Peer-to-peer diffusion (stage 3a)	Commercial diffusion (stage 3b)
Control variables				
Gender (male)	.39 (.11)***	.62 (.14)***	-.21 (.27)	.44 (.26)
Technical background	.34 (.11)**	-.05 (.13)	.29 (.21)	-.27 (.19)
Inspiring social environment	.49 (.12)***	.14 (.15)	.50 (.28)	-.02 (.21)
Frequency of unmet needs	.62 (.12)***	.61 (.15)***	-.10 (.25)	.30 (.21)
Big Five personality traits				
Openness to experience	.35 (.11)**	.08 (.14)	.21 (.24)	-.09 (.20)
Extraversion	.12 (.11)	-.51 (.16)**	-.28 (.27)	.12 (.22)
Conscientiousness	-.13 (.11)	.31 (.15)*	-.64 (.28)*	.57 (.28)*
Agreeableness	.03 (.11)	-.06 (.14)	-.40 (.25)	-.28 (.24)
Neuroticism	-.07 (.11)	-.13 (.15)	-.35 (.32)	.42 (.22)
Constants	.59 (.10)***	.13 (.13)	-1.89 (.30)***	-1.53 (.29)***
Model Fit				
Wald test statistic (degrees of freedom)		96.36 (9)***		

Source: Stock, von Hippel, and Gillert 2016, table 3. $n = 547$. Analysis method: sequential logit regression. Coefficients reported in log-odds units; robust standard errors in parentheses; degrees of freedom = 9. *$p < .05$, **$p < .01$, ***$p < .001$.

The control variable "inspiring social environment" was included because the social environment within which innovation takes place has been found to be important to innovation likelihood and success. An inspiring environment involves strong social ties (Perry-Smith 2006) and also supportive attitudes toward innovation (Amabile, Conti, Coon, Lazenby, and Herron 1996; Scott and Bruce 1994). For example, a supportive family would say to an ill family member attempting to innovate, "How wonderful that you are being creative in that way, how can we help?," as opposed to, "Why are you doing something so foolish? You should just follow your doctor's orders!" As can be seen from the third row of the table, this variable is significantly correlated with successful idea generation.

The fourth control variable, "frequency of unmet needs," refers to the degree to which a respondent felt that he or she had needs not satisfied by products on the market, and so would have a reason to innovate. The association of this variable with innovation likelihood has been documented in numerous studies of innovation by lead users (e.g., Morrison, Roberts, and Midgely 2004; Franke and von Hippel 2003). As can be seen in row four of table 9.1, this control variable was significantly associated with both successful completion of the idea generation phase and completion of the prototype phase too.

Findings regarding personality traits

In the bottom half of table 9.1 we see the personality traits significantly associated with successful completion of each stage in the innovation process. These differ significantly from stage to stage. In the first stage, table 9.1 shows that individuals high on "openness to experience" are significantly more likely to have an idea for an innovation. This makes sense: openness to experience has been consistently shown to positively affect creative behaviors for different groups of employees (Feist 1998; George and Zhou 2001; Rothmann and Coetzer 2003; Sung and Choi 2009; Wolfradt and Pretz 2001).

In the second stage, introversion (the negative of "extraversion") and "conscientiousness" are significantly associated with those who successfully create a prototype for personal use. A connection between introversion and "working on technical things in a lab" also fits prior research findings. Thus, in a study by Lounsbury et al. (2012), scientists were found to have significantly lower levels of extraversion than nonscientists. Similarly, Williamson, Lounsbury, and Han (2013) found that engineers scored lower on extraversion than non-engineers. With respect to conscientiousness, it is reasonable that those working on prototypes would have this trait. To my knowledge, this is a new finding.

In the diffusion phase, my colleagues and I distinguished between peer-to-peer diffusion and commercial diffusion. We did so because we thought that accomplishing each successfully involved very different activities and personality traits. This final phase is clearly important to innovation success and also to the social benefit derived from the free innovations created. However, because the individuals reaching this final phase had already had some personality traits strongly selected for, we really did not have enough variation left in the sample to see much in the way of significant associations between personality traits and successful diffusion. As table 9.1 shows, we did find a correlation at a low level of significance ($p < .05$) between the personality trait of conscientiousness and diffusion success. Individuals who possessed *high* levels of conscientiousness were more likely to succeed in diffusing their innovations commercially. In contrast, those who were *less* conscientious were more likely to succeed in peer-to-peer diffusion. An explanation for this pattern is not clear to me and, given the modest statistical significance of the findings, I will not attempt interpretation.

How Personality Traits Affect the Success of Free Innovation Projects

To understand the practical effects of personality traits on success in innovation, we calculated marginal effects at the means (MEM). This involved calculating the change in probabilities produced by a one-unit change on a seven-point Likert scale in a single personality trait variable, while holding all other trait variables at their mean values. MEM calculations do show that personality traits are important to free innovation project success.

The Big Five traits jointly accounted for 9.6 percent of variance in successful completion of the ideation stage (Nagelkerke 1991), where success was based largely on openness to experience. A MEM calculation shows that a one-unit increase in openness, with all other predictors held at their mean values, increased the probability of successfully completing the ideation stage by 9.5 percent. At the prototyping stage, the variance solely accounted by the Big Five was 8.0 percent. Being an introvert and being conscientious were both significantly associated with the likelihood of success in prototyping. Here MEM calculations show that a one-unit increase in extraversion decreased the probability of successful prototype completion by 15.1 percent, whereas increasing conscientiousness by one unit increased the probability of successfully completing a prototype by 9.7 percent.

If we next consider the *combination* of traits an individual must possess to successfully traverse the successive stages, the cumulative effect of personality traits on household innovator success becomes clear. As was mentioned earlier, personality traits significantly associated with successful completion of earlier stages are automatically carried into subsequent stages by the individuals who possess them. For example, as can be seen in table 9.1, the personality traits favorable to successfully completing the first two stages were openness, conscientiousness, and introversion. Individuals in our sample who were found to be at the "90 percent favorable value" for these three traits, with the remainder set at mean values, had a 52.9 percent chance of successfully completing both stages. For individuals displaying the combination of traits least associated with success—low openness to experience and low conscientiousness (tenth percentile) but high extraversion (ninetieth percentile)—the probability of successfully completing both stages was only 16.1 percent.

Discussion

We have seen that a number of factors can significantly affect innovators' likelihood of success at completing three basic stages in the process of developing and diffusing an innovation. In the main, the findings are intuitively very reasonable with respect to both the control variables we included in the study and personality traits. For example, it makes sense that individuals who have lots of unserved needs will more be likely to have ideas about how to solve them and thus succeed at the ideation phase of innovation. And it makes sense that if those unserved needs are strong, an individual will be more motivated to at least attempt to build a prototype, other things being equal. More generally, it makes sense that having the skills, resources, and personality traits appropriate to completing a certain stage of innovation will make successful completion of that stage more likely.

Is there a way to convert these findings into practical ways to increase levels of successful innovation in the household sector? At first glance things do not look too promising, as most of the control variables and the personality variables in table 9.1 are not easy to adjust. Consider that increasing one's level of technical education requires a major personal investment. Further, personality traits are essentially stable in adulthood. And, if one does not have an inspiring home environment, changing that will probably not be easy either.

However, my research colleagues and I think there are two accomplishable approaches that are likely to yield major benefits. First, one can encourage collaboration, so that people can help one another "fill in their personal gaps" in resources, training, or personalities. Second, one can use technical advances now available to free innovators to make innovation development and diffusion tasks both less demanding and less trait-specific.

Encouraging collaboration

Recall that today the dominant pattern in household sector innovation is that all innovation process steps are completed by a single individual. As table 2.6 documents, studies of household sector innovation in the United Kingdom, the United States, and Japan have shown that in those countries about 90 percent of innovations are made by individuals acting alone. In Finland and South Korea, 72 percent of innovations by consumers are made by individuals acting alone, with the remainder being collaborative efforts.

As was discussed earlier, an individual acting alone may be well prepared in terms of personality traits to succeed at one innovation stage but less well prepared for the next stage, where the identical traits are less helpful. When innovation is collaborative, it may be possible to solve this problem: *Collectively* the collaborators may have all the personality traits needed to successfully complete all three stages of innovation. A start-up firm uses this strategy when it puts different types of people on a team. When a new business venture is created to develop, produce, and market an innovation, it is a common prescription for success to recruit a group of individuals who *collectively* have expertise in all tasks relevant to the project (Akgün, Keskin, and Byrne 2010; Ensley and Hmieleski 2005; Vissers and Dankbaar 2002). The same strategy is often used by the personnel departments of larger firms (Muchinsky and Monahan 1987; Kristof 1996).

Innovations developed collaboratively also diffuse more frequently than do innovations developed by single individuals. The difference can be quite striking, as was noted in chapter 5. Thus, recall that Ogawa and Pongtanalert (2013) found that when individuals belonged to communities with a shared interest in the innovation they developed, the adoption rate by peers was 48.5 percent. When innovators did not belong to such communities, the adoption rate was only 13.3 percent. Other literature supports these patterns. For example, it is clear that innovators participating in communities tend to share information, including information about innovations they have developed, with other members (Morrison, Roberts, and von Hippel 2000; Raasch, Herstatt, and Lock 2008).

In view of the evidence of the benefits associated with collaborative innovation, policymakers and practitioners may wish to explore ways to increase the proportion of collaborative projects in the household sector. Increasing the availability of innovation facilities such as makerspaces is one potentially useful practical step. Such facilities offer access to sophisticated prototyping tools; they also enable potential collaborators to congregate and to discover one another. Also likely to be helpful are online community forums in which people can post their innovation-related interests and find one another at low cost. One excellent example of such a forum is https://patient-innovation. com/, a non-profit website that provides a collection point for information on patient-developed innovations (Patient Innovation 2016). That website is also designed to support online discussion and sharing of innovation-related information by medical patients and people interested in helping them (Habicht, Oliveira, and Shcherbatiuk 2012). More generally, of course, inexpensive Internet access and toolkits for collaborative design, such as those supplied by and for open source software development communities, can support collaboration at a distance.

Changing the nature of innovation tasks

A second approach, complementary to the first, is to change the nature of innovation tasks to ease the resource and personality attribute constraints required to accomplish them successfully. This approach, enabled by improvements in innovation development tools available to individuals, is becoming increasingly feasible.

Tools derived from creativity research, such as those supporting analogical thinking, are widely available today and can assist innovators in "thinking outside the box." These tools may help individuals (even those in whom openness is not a strong personality trait) to develop innovation-related ideas. Inexpensive CAD programs increasingly enable even novices to create robust designs more easily and quickly than they could in the past. Manual skills associated with prototyping, such as using saws, hammers, and glue, are giving way to computer-aided manufacturing. Computer-driven fabrication tools such as 3D printers make it practicable to produce parts for a prototype at the push of a button. These methods may reduce the importance of introversion and conscientiousness as personality traits associated with successful prototyping.

With respect to innovation diffusion, face-to-face describing or selling may be at least partly replaceable by a diffusion process based heavily on Internet postings. To non-extroverts, such a process may be more congenial than face-to-face diffusion tasks.

In sum, my colleagues and I conclude that several factors, significantly including personality traits, affect the success of household sector innovators. It seems likely that attention to those factors could enable societies to increase the number of innovation projects attempted by householders, and also to increase the fraction of their projects that progress to a successful conclusion.

10 | Preserving Free Innovators' Legal Rights

Rules and regulations are so pervasive in many countries that it is easy to assume that only professionals are allowed—or should be allowed—to innovate. Is it really safe to let just anyone innovate? Or, as my mother would sometimes urgently frame the question to my father: "Arthur, are you going to just stand there and let your son do *that*? He might blow up the house!" (Actually, I never did.)

It is true that innovation is not always risk-free and that many individuals and social institutions are risk averse. So it is fortunate for us all that individuals, especially in common law democracies like the United States, the United Kingdom, and Canada, have broad legal rights to develop and use innovations.

In this chapter, drawing on work by Torrance and von Hippel (2015), I review the fundamental legal rights of household sector innovators, including free innovators. I then describe how governments can and do encroach on those important rights—often without intent, and in pursuit of other objectives. Andrew Torrance and I conclude that a strengthened social awareness of the need to protect individuals' rights to innovate would be very valuable. We suggest how this might be accomplished.

Individual Innovators' Legal Rights

In the United States, individuals have fundamental legal rights to engage in free innovation development, to use their innovations, and to publicly disclose and discuss them. These rights are embedded in both the common law and the United States Constitution (Torrance and von Hippel 2015).

The common law is a body of legal principles that continuously evolves from customary practices and the decisions of courts. A fundamental principle of the common law that supports individuals' rights to innovate is the principle of "bounded liberty": that in the absence of specific and legitimate prohibitions, people are at liberty to act however they choose. However, that liberty is bounded in the sense that people are at the same time restricted from taking actions that materially harm others. Thomas Jefferson (1819) stated that "rightful liberty is unobstructed action according to our will, within limits drawn around us by the equal rights of others." Later, the First Amendment scholar Zechariah Chafee, Jr. (1919, 957) stated the same idea more vividly as "[the] right to swing your arms ends where the other man's nose begins."

With respect to innovation, the common law principle of bounded liberty informs us that individuals have a right to engage in innovation without needing permission from other people or from governmental entities provided that their actions are not unreasonably dangerous to others and do not violate specific and legitimate legal prohibitions.

Individual innovators are also shielded by robust rights to privacy derived from common law, statutes, and, in the United States, the United States Constitution. This right to privacy provides formidable protection against intrusion, particularly governmental intrusion. It enables people to innovate in privacy in ways that might be controversial if known, and to go through early learnings and failures protected from immediate public scrutiny. In his classic textbook on tort law Thomas Cooley (1879, 29) provided an early description of a common law right of personal autonomy: "The right to one's person may be said to be a right of complete immunity: to be let alone." Later, the legal scholars Samuel Warren and Louis Brandeis (1890) formally proposed, and helped to establish the existence of a constitutional right to privacy.

Individuals in the United States also have robust rights to innovate *collaboratively* and to diffuse information about their innovations to others openly. These rights are guaranteed by the First Amendment to the Constitution, which states in Article 1 that "Congress shall make no law ... abridging the freedom of speech, or of the press; of the right of the people peaceably to assemble." That amendment, through incorporation by the Fourteenth Amendment, also prohibits state or local governments from creating laws abridging freedom of speech. Protected by these rights, free innovators can get together physically or virtually, and can collaborate by exchanging information on work in progress. They can also diffuse their designs and their observations regarding their functioning to any and all, absent a compelling governmental need such as national security.

Taken together, the legal rights just described create a powerful shield for those wishing to pursue free innovation either singly or collaboratively and to diffuse their designs and findings freely and widely.

How Legislation and Regulation Affect Free Innovation

In view of the array of legal rights just described, one might ask why individuals' rights to develop and apply free innovations are not secure. Again, recall Chafee's rule: "[the] right to swing your arms ends where the other man's nose begins." The issue then is that the development and the use of some innovations may be sources of harm to public or private interests. This can create a reasonable basis in law and policy to constrain individual innovators' liberty of action when these conditions hold.

In the United States, federal, state, and local governments can affect individuals' rights to innovate. Each of those levels of government can constrain or support consumers' freedoms to innovate by means of court decisions, statutes, regulations, or even informal policies intended to promote public safety, welfare, and property rights (including intellectual property rights), among other motivations. Constraints can be direct (as when building codes restrict novel building techniques in the name of safety) and/or indirect (as when development and practice of innovations require access to a public resource). Consider that one can build any type of car one likes, but that to test it or use it on public roads one must meet detailed regulatory requirements intended to protect the safety of others. Similarly, in the US, one can build a drone aircraft, but to test or use it in the public airspace one must adhere to detailed regulations set forth by the Federal Aviation Administration (FAA) or risk severe penalties. Also similarly, in the US one can build a new wireless transmitter, but to test or to use it on the public radio spectrum one must adhere to regulations set forth by the Federal Communications Commission (FCC).

Federal, state, and local legislative and regulatory bodies can and do take actions that raise the costs of or otherwise restrict free innovation by individuals as they pursue their mandates to promote public safety, welfare, or other aspects of the public interest. Often, Torrance and I find, legislators and regulators negatively impact free innovation without intention or even awareness, simply as a side effect of regulations promulgated for other purposes, such as regulation of industry.

As an example of legislation that has raised the cost of a large amount of free innovation, apparently without that intention, consider the Digital Millennium Copyright Act of 1998 (DMCA 1998). That federal act was intended to prevent free digital copying—"piracy"—of copyrighted and commercially sold information products such as software and music. Specifically, the DMCA made it a crime to circumvent the anti-piracy measures built into many digital products. The intent of the law was to reduce digital piracy. The thinking was that, if an individual was, at pain of criminal sanctions, prevented from gaining access to the code, that individual would be prevented from creating pirate copies.

Because of the approach that was taken, the DMCA has caused severe "collateral damage" to free innovators' abilities to innovate utilizing software-containing products that they have purchased legally. The problem is both free and commercial innovators need access to software code in order to understand, modify, and improve products they purchase. Absent the DMCA, these activities would surely otherwise be legal as "fair use" (also known as fair dealing) exceptions from copyright infringement. In effect, while intending to combat digital piracy, the DMCA raised the costs of some types of free innovation significantly, and even denied innovators access to some of their own recognized legal rights (Electronic Frontier Foundation 2013).

The damage inflicted on free innovation by the DMCA is not quantifiable—no one can total up the value of projects not embarked upon—but it may well be significant in scale. Recall that, in a survey that was discussed in chapter 2, 14 percent of innovation by users in the United Kingdom involved the development and the modification of software. If in the United States the same percentage of innovation is devoted to software (something that was not measured in the US survey), an annual total of $2.8 billion of valuable user innovation activity in the US alone has been put at some level of risk by the DMCA. Again, in this instance the damage to free innovation was apparently unintended.

Torrance (2015) finds no evidence that the drafters of the DMCA were even aware of free innovation, much less of the damage their legislation might inflict upon it, although they did express some concerns about the loss of fair use rights.

The Relative Advantage of Free Innovators

Despite restrictions like the DMCA, free innovators can have an advantage over producer innovators—they may have greater "freedom to operate" with respect to both regulations and the law. First, consider the practical matter that free innovation, relative to producer innovation, is often small in scale, is widely distributed, and may be practiced in the privacy of one's home. As a result, possible violations of regulation and law on the part of free innovators are as a practical matter not easily discovered. For example, if a free innovator develops and builds and uses an innovation that draws upon a patented invention without permission (and often without awareness), that infraction is likely to escape notice. Second, it is the case that the common law in the US has a principle of *de minimus*—a principle of ignoring very small violations. This legal principle, too, gives a systematic advantage to free innovators relative to larger-scale producers.

There is also a very important source of advantage held by free innovators in the US that does not depend upon small scale—it is built into the US Constitution. In the United States, federal regulatory agencies are restricted to regulating *commercial* "interstate commerce" only. Specifically, the power of federal agencies to regulate is largely derived from the Commerce Clause in Article 1, Section 8, Clause 3, of the US Constitution (*Constitution*). This clause grants Congress power "to regulate Commerce with foreign Nations, and among the several States, and with the Indian Tribes." The Supreme Court has construed the Commerce Clause as permitting Congress to pass statutes regulating *commercial* activity that implicates interstate commerce either directly or indirectly. The full extent of this legal authority has waxed and waned over the years in concert with Supreme Court philosophy. However, Supreme Court decisions have consistently agreed that the Commerce Clause does not allow federal agencies to regulate truly noncommercial activities. The Supreme Court reaffirmed this principle in 2012 when it decided *National Federation of Independent Business v. Sebelius*. There, the Court clarified that "the power to *regulate* commerce presupposes the existence of commercial activity to be regulated" (National Federation 2012, 18).

Differential regulatory treatment of free innovators vs. producers arising from the Commerce Clause can greatly advantage free innovators, especially in fields that are highly regulated, like medical treatments and devices. For example, free innovators can develop and use their own medical drugs and devices, perhaps the very same ones that highly regulated producers are striving to get governmental approvals to market, entirely free of regulatory oversight by the Food and Drug Administration (FDA)—*if* they do so noncommercially. Free innovators are also free to distribute information about their innovations, including design details and the effects of use they have experienced, to others without permission from, or constraint by, the FDA or the Federal Trade Commission, as long they do this for free, and do not implicate governmental interests vital enough to allow abridgement of their free speech rights. Additional individuals can then make noncommercial copies for themselves, and are free to personally use these without FDA control or oversight. Of course, legal constraints to these activities may still exist apart from federal regulations.

Producers that develop and *sell* novel drugs or devices or services for medical use are of course in an entirely different situation. Participating in commerce, such as by selling, triggers the Commerce Clause, and gives the FDA and other relevant agencies jurisdiction to regulate. In highly regulated fields, where innovation by producers is made especially costly, the net result may be that the pattern of free innovator, grass-roots pioneering described earlier is highly advantaged relative to producer innovation, and may become very strong indeed.

Possible Legislative and Regulatory Improvements

We have now seen that free innovators—both those acting individually and those acting collaboratively—have strong and fundamental legal rights to develop and diffuse innovations noncommercially. Indeed, at least with respect to federal regulation, free innovation projects operate with fewer legal constraints than do producer innovation projects.

With increased awareness of the potential benefits of free innovation, creative legislation and policymaking can be used to further expand the benefits of the free innovation paradigm.

In one generic approach, agencies can elect to open up segments of public resources for unlicensed use and experimentation by free innovators and commercial innovators. For example, the US Federal Communications Commission reserves some segments of the radio spectrum as "white space" in which individuals or firms can explore and exploit novel uses without a license (Barnouw 1966; FCC 2015). This policy approach can yield great benefits. For example, many of the successes of unlicensed wireless, like the development and extension of the range of WiFi, have been developed in these unlicensed spaces by free innovators and commercial firms alike (Sandvig 2012). At the same time, Congress and the FCC reserve other parts of the radio spectrum for exclusive use by specific regulated uses, such as on-air TV stations. Similarly, the US Federal Aviation Administration allows the use of some airspace—e.g., space far from airports and up to a height of 400 feet within visible range—for unlicensed and noncommercial use by hobbyists who build and operate small radio-controlled model airplanes and drones. Other altitudes and areas are reserved for the use of pilots of licensed aircraft or are completely off limits to use by free innovators.

A second generic approach is to settle upon a more generous and generative interpretation of the organic statutes governing agencies, and in this way shift agency regulation into a posture more friendly to free innovation. For example, part of the statutory mission of the FDA is to "promote the public health" (21 U.S. Code § 393(b)(1)) and to "protect the public health" (21 U.S. Code § 393(b)(2)) with respect to foods, drugs, medical devices, and the like. Rather than interpreting this as a mandate for restricting innovations, the agency could decide that its mission could be better accomplished by being agnostic to or even promoting free innovation.

This more generative regulatory approach can be applied by any regulatory body, as is illustrated in Section 104.11 of the International Building Code (Alternate Materials, Design and Methods of Construction, IBC 2009). This Code Section, used in Utah and in some other states, gives county building inspectors flexibility in approving the use of unconventional building materials. Instead of approving only specified materials, as is common in many building codes, inspectors may approve any material as long as they are satisfied that it meets the functional requirements of safety and reliability. Such a regulation has notable advantages. It allows for innovation in building materials, which may lead to improved materials, but it also maintains sound public policy by ensuring that the materials are safe and work as intended (Harris 2012). Similar flexible treatment of free innovation can be found in regulations applied to experimental airplanes by the Federal Aviation Administration.

In a third generic approach, the federal government can insist that already mandated cost-benefit analyses of proposed federal regulatory actions include assessments of effects on free innovation. The Reagan administration was the first to make cost-benefit analysis a requirement for all federal regulatory agencies. On February 17, 1981, President Ronald Reagan promulgated an executive order that mandated cost-benefit analyses of federal regulations when triggered by any of a variety of factors, most of them economic. Among the triggers was any rule "likely to result in ... significant adverse effects on ... innovation" (Executive Order 12,291). Succeeding presidents largely maintained this approach. On January 18, 2011, President Barack Obama issued an executive order that stated that "each agency shall ... seek to identify, as appropriate, means to achieve regulatory goals that are designed to promote innovation" and that the effects of past as well as future regulations are subject to assessment (Executive Order 13563).

Applying cost-benefit analysis to possible effects on free innovation is becoming increasingly practicable as measurement of free innovation improves. As was noted earlier in this chapter, Torrance and I were able to quantify roughly the extent to which free innovation in software development in the United States could be adversely affected by the Digital Millennium Copyright Act. Once negative effects on free innovation have been shown, there will be a basis for adjusting specific laws and regulations found to have deleterious effects. Thus, legislators could amend the DMCA to ensure that free innovators' traditional rights to reverse engineer and improve products that they purchase are no longer encumbered (Stoltz 2015).

As a second example, consider that intellectual property rights may have negative effects on free innovation. By definition, free innovators do not themselves acquire intellectual property rights. However, rights held by others can reduce free innovators' freedom to operate because present law does not provide a "home use" or noncommercial exemption for free innovators. To correct this, Congress could pass legislation exempting individuals from liability for copying patented inventions for personal and noncommercial use or for experimental use. Home use exemptions already are in place in other countries. Allowing such exemptions in the United States would eliminate a cost-raising risk for free innovators. Benkler (2016) explains in detail how, within US law, an expanded "experimental use exemption" could be effective if implemented along with related changes. If done judiciously, changes such as those he suggests would have only a negligible effect on producers' incentives to innovate.

Discussion

Through the research and the discussions presented in this book, my colleagues and I have argued and shown that free innovation is very generally valuable for individual innovators, for producers' profits, and for social welfare. Solidifying legal, regulatory, and social support for free innovation will require an increase in general awareness of free innovation and of the benefits it provides. In our 2015 paper, Torrance and I suggest that a term that may be useful in that regard: "innovation wetlands." Just as Boyle (1997) did with respect to popularizing the value of the intellectual commons, we draw an analogy to successful efforts made in recent decades to create general awareness of the great public benefits that environmental wetlands provide.

Consider that until the 1970s marshy ecosystems were generally regarded as, at best, resources ripe for conversion to more beneficial uses. At worst, they were considered noxious sources of pestilence and disease, as exemplified by the disparaging phrase "malarial swamp." Accordingly, for many decades governments promoted the filling or the draining of wetlands through a variety of legislative and policy instruments. For example, the Watershed Protection and Flood Prevention Act (1954) directly and indirectly increased the drainage of wetlands near flood-control projects. Tile drainage and open-ditch drainage were considered conservation practices under the Agriculture Conservation Program. These and other policies caused losses of wetlands averaging 550,000 acres a year from the mid 1950s to the mid 1970s (Dahl and Allord 1997).

Beginning in the 1950s, a paradigm shift in scientific understanding of wetland ecology drove the recognition that, far from being dangerous or waste areas, wetlands are actually among the most productive and diverse of ecosystems, providing such benefits as habitats for diverse species, flood control, and water purification. Diffusion of information about these benefits changed society's perception of wetlands, and the posture of governments also changed. "Noxious swamps" increasingly came to be viewed as "valuable wetlands." Changes in regulatory approaches resulted in in a new emphasis on protection, preservation, and even rehabilitation of degraded wetlands both nationally and internationally (Clean Water Act 1972; Ramsar Convention 1971). Where governments had once targeted wetlands for destruction, many now focus on preserving them.

Torrance and I define the *"innovation* wetlands" as the rights and conditions that enable free innovation by individuals to flourish. Just as is true of environmental wetlands, the nature and the extent of innovation wetlands must be understood, and the value of the innovation activity that takes place within them must be better appreciated.

A more enlightened understanding of the benefits of free innovation can create a climate within which regulators and firms will be able to work with free innovation rather than against it. As illustration, consider again the very interesting example of medical patients' freedom to innovate relative to highly regulated medical producers. In line with the general case discussed in chapter 6, free innovator pioneering may turn out to be a very good thing for rapid medical progress and for medical producer firms—*especially* if public understanding enables intelligent support from producers and regulators and legislators rather than resistance. As we saw, free innovators have the *right* to innovate to help themselves, and are clearly impatient to do so. The motto of the Nightscout free innovator group that develops medical devices for diabetics (discussed in chapter 1) is "#WeAreNotWaiting" (Owen 2015; Nightscout project 2016). By this, the Nightscout group is saying that it rejects the common pattern of producer and FDA instructions to wait for promised commercial solutions to their urgent medical needs—commercial solutions that always seem to be five years away. And, indeed, why should patients wait for commercial solutions when they can instead effectively innovate to help themselves?

Medical self-experimentation clearly has dangers to the individuals who, nonetheless, have a *right* to risk danger in order to help themselves. There will clearly be instances of failure and injury or even death from such experimentation. *But* there will also predictably be great progress, including life-saving help for many. A climate of understanding and support for the overall value of the enterprise will enable legislators and regulators to resist "clamping down" in response to specific unfortunate failures. Instead, they will be able to offer intelligent support to enable free innovators to innovate more safely, and to better assess the actual safety and efficiency of the innovations they develop.

As an example, consider that today the US Food and Drug Administration, along with governmental agencies of similar function around the world, supports a "gold standard" system of clinical trials. This system has evolved over time, and today has become so expensive that it is viable only for drug and device innovations that offer the potential for very high profits. Many very important and also commonplace medical innovations have no chance of getting evaluations via clinical trials under this system. For example, a new device or method for assisting getting out of bed in the morning may be very valuable to many disabled or elderly individuals—but it would not be cost-effective to test its efficacy and safety via clinical trials of the type mandated by the FDA.

Rather than attempting to suppress the development, personal use, or diffusion of free innovators' medical innovations, public awareness of the value of that activity could support more positive responses. For example, public and producer support could help to develop user-friendly, affordable clinical trial methods that enable free innovator communities to quickly evaluate the efficacy and safety of free innovations. The practicality of patient-run clinical trials has been demonstrated, for example, in a clinical trial of possible ALS therapies (Wicks, Vaughan, Massagli, and Heywood 2011; see also DoubleBlinded

2016). At least initially, of course, the methodologies of these community-based trials will not be at the level of the FDA gold standard. But the FDA gold standard also was not built in a day. With public understanding, the FDA, producers, and legislators would be enabled to support the development of a grass-roots complement to the FDA system that will steadily improve over time.

Without public support, in contrast, FDA regulators might be motivated or even forced to attempt to suppress free innovation even in the face of Commerce Clause restrictions. Citing, for example, the possibility that malevolent individuals might "hack" medical devices, the FDA could try to make free innovation more costly. It could, for example, compel producers of medical devices—firms that *are* under the purview of the FDA—to make the devices they sell more difficult for medical patients to reverse engineer, extract personal data from, and otherwise improve for their own use. (For example, the NightScout innovators described in chapter 1 did require access to the personal medical measurements generated by commercial medical devices as inputs to their free and very useful designs.) The net result would, in my view, produce damage very similar to that caused by the overbroad legislative response to digital piracy via the DMCA discussed previously. Alternative responses that prevent malevolent hacking but at the same time grant "owner override" to owners and users who wish to modify their own devices and systems are both possible and, in my view, clearly preferable (Schoen 2003).

Prominent legal scholars both support and urge such a transition in public thinking. Pamela Samuelson (2015, 1) explains the importance of "freedom to tinker" in well-protected innovation wetlands, noting that "people tinker with technologies and other human-made artifacts for a variety of reasons: to have fun, to be playful, to learn how things work, to discern their flaws or vulnerabilities, to build their skills, to become more actualized, to repair or make improvements to the artifacts, to adapt them to new purposes, and occasionally, to be destructive." She urges efforts to preserve and legally protect the freedom to tinker. William W. Fisher III (2010) similarly argues that creative tinkering is fundamentally important to "human flourishing" and summarizes psychological and philosophical research to argue that user innovation is an important pathway to self-fulfillment, the richly lived life, and human happiness.

In net, Torrance and I conclude that free innovation is important to both human happiness and inventive progress. We find that fundamental legal protections afforded to free innovators are robust. At the same time we argue, along with colleagues, that better stewardship of the innovation wetlands can be created by greater public and governmental understanding of the beneficial effects free innovation brings to individuals, to social welfare, and to national economies.

11 | Next Steps for Free Innovation Research and Practice

Free innovation is, as we have seen, an important and growing "grass-roots" innovation process in the household sector of national economies. Free from compensated transactions, it is fundamentally simpler than producer innovation. In this concluding chapter, I propose some specific next steps for those interested in further work on the theory, the policy, and the practice of free innovation. Of course my list of suggestions is simply that: others will certainly have many other excellent ideas.

Proposed Next Steps

As we have seen in this book, the free innovation paradigm provides a novel and generative framework for understanding innovation in the household sector. It, together with the producer innovation paradigm, offers novel and expanded space for innovation theory development, empirical research, policymaking, and practice. In the sections that follow, I discuss some issues and possible new lines of inquiry in each of these domains. In addition, with respect to valuable next steps, readers should note that research on all aspects of the free innovation paradigm is at an early stage. They should therefore consider the theoretical and empirical work presented in each of the preceding chapters as both inviting and requiring further development.

I begin by comparing the research lenses offered by the concepts of free innovation, user innovation, peer production, and open innovation. Researchers of course have a choice of conceptual lenses for their studies, with each most suited to some topics and styles of inquiry. I then focus in on issues and questions related to free innovation only. I begin by proposing steps to improve the measurement of free innovation, a matter that is very important to further progress. I next suggest steps to incorporate free innovation into microeconomic theory, and also into important components of innovation policy. Then, I suggest how the free innovation paradigm can help us to understand the economics of both open source producer innovation activities, and household sector creative activities beyond innovation, such as "user-generated content" ranging from fan fiction to contributions to Wikipedia.

Finally, I conclude the book by again proposing that it will be very important to seek to understand free innovation and the free innovation paradigm more deeply via further research. Free innovation offers the promise of empowering all of us in the "household sector"—simultaneously enriching our individual lives, increasing social welfare, and improving national economies.

Free Innovation and Related Research "Lenses"

Again recall from chapter 1 that I define a free innovation as one that (1) is developed by consumers at private cost during their unpaid discretionary time, and (2) is not protected by its developers, and so is potentially acquirable by anyone for free. This definition of free innovation is intentionally very restrictive. It dictates that free innovation models and samples must contain *no* compensated transactions of any kind, and that innovation development work be entirely self-rewarded. The purpose of creating this very precise and tight lens is that it excludes many potentially interfering variables, and so enables researchers to more clearly analyze phenomena central to free innovation. Illustrative examples of such phenomena discussed earlier are innovation pioneering by free innovators, and the likely dearth among free innovators of incentives to diffuse.

Of course, much of the world is hybrid, and so will diverge from our definition of free innovation to a smaller or larger extent. This is an opportunity rather than a problem for researchers. One can first isolate and analyze interesting phenomena within the precise lens of free innovation, and then progressively relax constraints in order to draw hybrid cases into consideration. Via progressive relaxations, one can learn whether and to what extent behaviors characteristic of free innovation endure in hybrids—and whether new behaviors emerge. For example, in some open source software and hardware projects today, contributors are exclusively free household sector innovators. In other projects, some or many contributors are paid by producers to participate. We may use this source of variation to explore the extent to which the addition of paid employees and related producer incentives changes the nature of open source projects and their outputs, with related gains or losses in social welfare.

Other research lenses that bring different aspects of the phenomenon of "non-producer innovation" into clear focus include commons-based peer production, user innovation, and open innovation. Researchers will wish to choose among these concepts and others—or to develop their own—as a function of the study question they address and the focus they prefer.

Commons-based peer production is term coined and brought to research prominence by Benkler (2002, 2006). It describes distributed "production" networks in which large numbers of contributors bring their own resources to an activity. They then work cooperatively, often via the Internet, to generate valuable outputs and reveal them to the commons.

Commons-based peer production shares many elements with free innovation. The most important distinction lies in the parsimony vs. inclusiveness of the two concepts. As I mentioned above, the free innovation lens is tightly constrained. In contrast, the commons-based peer production framework incorporates much more richness and complexity. Thus, while free innovators must be self-rewarding, participants in commons-based peer production need not be: contributors to peer production projects may be either self-rewarding or paid for their work. Similarly, free innovators must not engage in compensated transactions during the course of innovation development and diffusion. In contrast, participants in peer production projects may engage in social and/or monetary transactions, and so incur related transaction costs. As a consequence of its inclusiveness, the commons-based peer production lens can be especially useful for richly descriptive studies of complex real-world situations. For the same reason, application of this lens can make quantitative analysis and modeling more difficult.

User innovation is sharply focused on the functional relationship that innovators have to an innovation they develop. If the innovator develops an innovation for personal or in-house use, he, she, or it is a user innovator. If the innovator develops the innovation to sell, he, she, or it is a producer innovator (von Hippel 1976, 1988, 2005). The presence or absence of self-rewards and compensated transactions does not play a role in this simple definition. As a consequence, the user innovation lens can include both free innovators and profit-seeking individuals and firms as user innovators. A user innovator firm, for example, would be one that develops a novel process machine for in-house use rather than sale. The firm is indeed a user—but, unlike free innovators, it is also seeking profit from using that machine in its operations.

The user innovation research lens is useful to distinguish between innovators who have first-hand vs. second-hand information regarding needs for a given innovation. Users, whether free innovators or firms, are the generators of need information. In contrast, producers must acquire it, with greater or lesser loss of fidelity, from users. This clear distinction, along with the concept of sticky information (von Hippel 1994), then allows us to understand why users and producers will have different local stocks of sticky information, and so will tend to develop different types of innovations. As a second matter, users, whether individual free innovators or user firms, are likely to care only about their own needs for an innovation, while producers, motivated by sales, must care about broader markets. This distinction can encourage innovation pioneering among all user innovators, as was documented in chapter 4 in the case of free household sector innovators.

Open innovation (Chesbrough 2003) falls squarely within the producer innovation paradigm. This lens is useful to explore and explain why and when a corporate strategy of both acquiring and selling innovative content and intellectual property can increase profits relative to a strategy of relying only on internally developed intellectual property. The term "open" refers to an *organizationally* open producer innovation process rather than to one involving an information commons. In that way open innovation is closely akin to the concept of technology marketplaces (Arora, Fosfori, and Gambardella 2001; Rivette and Kline 1999). With respect to research questions related to free innovation, the open innovation lens can be useful for exploring producer strategies for profitably linking to that phenomenon.

Measure Free Innovation

In this book, I have striven to characterize and explore free innovation in ways compatible with economic theorizing and analysis. I have done this even though free innovation clearly is not fundamentally or even mostly "about money." Instead, as studies of free innovators' motivations have shown, free innovation is most directly "about" a wide range of human interests and values having to do with utility, participation, fun, learning, creativity, altruism, and other important matters associated with "human flourishing" (Fisher 2010; W. von Hippel, Hayward, Baker, Dubbs, and E. von Hippel 2016). Still, in order to conduct analyses that can apply to activities in both the free innovation paradigm and the producer innovation paradigm, economic measures common to both are needed.

Developing ways to measure free innovation that are compatible with economic analysis is not a straightforward task. In free innovation, in sharp contrast with producer innovation, there are no transactions that can be used to document the value of investments made, and outputs created and diffused. Also, in free innovation there are no equivalents to patents as markers of development originality since free innovators do not apply for patents. Still, compatible measures of activities within the two paradigms and paradigm outputs can be devised. In view of the extent and importance of free innovation, work toward this end clearly will be worth the effort. Attempts to assign value to unpriced product flows have already begun, and improvements will doubtless follow. (See, e.g., Brynjolfsson and Oh 2012; Ghosh 1998.)

At present, household sector free innovation is *not measured at all* in official governmental statistics. In part this is because, in line with the traditional Schumpeterian producer-centric assumptions, official efforts to collect data on innovation are largely focused on enterprises in the business sector. In additional part it is because innovations developed by free innovators and made available for free diffusion do not fit the present-day official OECD definition for innovations. Recall from chapter 1 that, within the OECD: "A common feature of an innovation is that it must have been *implemented*. A new or improved product is implemented when it is introduced on the market" (*Oslo Manual* 2005, paragraph 150). Of course, free innovations are *not* diffused via the market—they are diffused for free and so are not *implemented* in OECD terms. Efforts to correct this problem by revising the official definition of innovation to incorporate a wider range of Internet-enabled diffusion options are needed (Gault 2012).

As long as it exists, the OECD's "on the market" requirement produces major distortions in measurements of innovation. Most directly, it hides free innovations generated in the household sector from view, because they do not fit the official definition of an innovation. Secondarily, it means that free innovations appear in official innovation statistics only if and when producers commercialize them. And at that point they are credited to *producers* as "new products introduced to the market" rather than to their actual, free innovator developers. This clearly misrepresents the sources of innovation. It also results in an overstatement of the productivity of producers' R&D for consumer products and services. The overstatement is likely to be substantial—several empirical studies have found that from about 50 percent to about 90 percent of major consumer innovations commercialized by producers were in fact initially developed by household sector innovators (Shah 2000; Hienerth, von Hippel, and Jensen 2014; Oliveira and von Hippel 2011; van der Boor, Oliveira, and Veloso 2014).

To date, and in the absence of official statistics collected by governments, statistics on free innovation have been collected by ad hoc empirical studies such as those discussed in this book. What is needed in addition, of course, is collection of rich data on innovation in the household sector on a regular basis. This will enable researchers to accumulate time-series data needed for many research purposes, ranging from studies of how free innovation is evolving, to studies of how it is affected by various conditions and interventions.

Social surveys of household sector innovators and surveys of producers both have a role in collecting the information needed for such work. Social surveys can be used to directly ask individuals in the household sector about their free innovation and their entrepreneurial innovation activities, the inputs they expended, and the outputs they created.

Social surveys can also be used to collect the "free innovators' side of the story" with respect to any diffusion of their innovations to both peers and producers. To get producers' complementary side of the story, governmental surveys of enterprises can be modified to ask about the incidence of and the value of adopting designs from free innovators. Initial experiments in this direction have been conducted by adding experimental questions to Community Innovation Surveys (CIS) in both Finland and Switzerland. These experiments demonstrate that valuable information on free innovation can be collected via the CIS.

Specifically, responses to the experimental questions added to the Finland CIS have shown that producers do indeed adopt customer designs as the basis for new commercial products, and that this can be important for their success in the marketplace (Kuusisto, Niemi, and Gault 2014). Some 6.1 percent of Finnish firms focused on consumer products in the 2014 Finnish CIS report that totally new product designs by end consumers are of medium or high importance to their product development. In addition, 8.7 percent of those firms report that modifications to their products by end consumers are of medium and high importance to them (Statistics Finland 2016, appendix tables 6 and 7). Analysis of the Swiss CIS experimental question findings further document the advantages to producers of a division of labor between free innovators and producers (Wörter, Trantopoulos, von Hippel, and von Krogh 2016.)

Incorporate Free Innovation into Microeconomic Theory

Despite the large and growing importance of free innovation in the household sector, free innovation has not yet been incorporated into standard microeconomic thinking. In part this is because statistical data series on free innovation do not exist yet. In part it is because, absent compelling data or other reasons, researchers with an interest in innovation may be quite satisfied to work within the traditional producer innovation paradigm, ignoring the important and growing levels of innovation in the household sector of national economies. After all, the Schumpeterian framework does fit a substantial portion of innovation development activity. Further, scholarly findings and data accumulated over many decades have made the producer paradigm an ever richer and more convivial environment for the conduct of normal science.

Expanding innovation research and research questions to include the free innovation paradigm offers very interesting new spaces for novel and enriched economic theories of innovation. Several illustrative examples have been initially explored in this book. In chapter 4, I explained why free innovators tend to pioneer new applications and markets, with producers following later. In chapter 5, I explored a market failure likely to reduce the diffusion of free innovations. In chapters 3 and 6, I discussed the potentially fruitful concept of a division of labor between free innovators and producer innovators. My colleagues and I have also shown that free innovation has positive effects on social welfare, and generally also on producers' profits, relative to a world in which only producers innovate.

Strikingly from the research perspective, my colleagues and I have documented that innovation activities in the free innovation paradigm do not require intellectual property rights to be viable. This finding can open the way to rethinking a central feature of microeconomic models of innovation: the assumption that private investments in innovation must be protected by systems of intellectual property rights. The argument underlying this assumption is that producers' profits from innovation investments will disappear if anyone can simply copy their innovations, and so producers must be granted exclusive control over their innovations for some period of time. (See Machlup and Penrose 1950; Teece 1986; Gallini and Scotchmer 2002.)

We now see that, even if producers do require intellectual property rights to protect and profit from their *own* investments in innovation design, adoption of free designs from free innovators requires much less producer investment—and so perhaps much less protection, too. This would be a welcome option to explore because, as is well known, intellectual property rights are a devil's bargain from society's point of view. At the same time as they (putatively) enhance producers' incentives to innovate, they also create deadweight losses for society by enabling monopoly pricing. Patents also disrupt the efficient forward movement of fields as owners of intellectual property place tollbooths astride promising pathways to further research and development (Murray and Stern 2007; Bessen and Maskin 2009; Murray, Aghion,

Dewatripont, Kolev, and Stern 2009; Dosi, Marengo, and Pasquali 2006; Merges and Nelson 1994). Efforts to ease these negative effects have a long history (e.g., Hall and Harhoff 2004). However, the inbuilt conflicts between social goals and producer goals with respect to intellectual property are fundamental, and problems will predictably fester.

A rethinking of the need for and effects of intellectual property rights should be based on an improved empirical understanding of where such rights are actually effective today. Sometimes patent rights do not exist in practice even when legally granted. Thus, biomedical researchers in universities and governmental and nonprofit institutions have been found to routinely ignore the legal rights of patent holders whose claims might impede their research (Walsh, Cho, and Cohen 2005). Contrastingly, many innovation types that are not legally protectable, and so assumed by economists to be freely available, are actually protected from potential free adopters by social means rather than legal means. For example, accomplished chefs cannot legally protect exclusive rights to the novel and economically important recipes they develop and practice in public—recipes are not patentable or copyrightable subject matter. However, these recipes are effectively protected nonetheless by community enforcement of anti-copying norms within communities of expert chefs (Fauchart and von Hippel 2008; King and Verona 2014).

Novel Policymaking for Free Innovation

The basic justification for public policy interventions to support innovation is to increase social welfare. Gambardella, Raasch, and I (2016) have made the case that social welfare increases when there is a division of labor between free innovation and producer innovation. Novel policies related to both development and diffusion of free innovations could be useful to support a transition to this improved condition.

Clearly, policy initiatives to support free innovation can include measures to reduce free innovators' development costs. These could include public funding of the development of open standards for the exchange of design information among free developers. Also, and analogous to the R&D subsidies provided to producers by government, support could be given to upgrading physical facilities used by free innovators, such as makerspaces (also sometimes called fab labs or hackerspaces) equipped with sophisticated tools that are beyond the means of most individual free innovators (Svennson and Hartmann 2016). Other infrastructure improvements could include support for the development of "big data" methods to identify, collect, and organize open public data on consumers' unmet needs. The net result would likely be an increase in both the number and the average social value of innovation opportunities worked upon by free innovators.

Recall from chapter 5 that free innovators are unlikely to have incentives to invest sufficiently in diffusing their innovations for free. Policy initiatives to support and reduce the costs to free innovators of the diffusion of their designs might help to reduce this investment shortfall. For example, free, easy-to-use public repositories of design information could serve this purpose. Such repositories should feature open documentation standards. In the absence of a strong push for open standards, proprietary repositories of free design information are likely to emerge, each tied to the proprietary standards of the sponsoring producer.

Gambardella, Raasch, and von Hippel (2016) explain that policy measures supporting *producers'* investments in supporting innovation development by free innovations should be designed to distinguish carefully between investments that complement free innovation and those that substitute for it. Public incentives for corporate R&D unambiguously raise welfare if they induce firms to invest in activities that are synergistic with free innovation. However, if public incentives instead support producer R&D that substitutes for innovative work that free innovation would do, the net effect can be to redistribute welfare from free innovators to firms, and perhaps also to lower aggregate social welfare.

Viable opportunities for free innovators are continuously increasing, due to technological trends that have been discussed. Accordingly, the appropriate division of labor between free innovators and producer innovators must continuously be updated. As an illustration, consider that patients and clinicians, during the course of regular medical practice, regularly discover new applications for drugs no longer under patent (DeMonaco, Ali, and von Hippel 2006). Producers, very reasonably, see no profit in investing in clinical trials to document the effectiveness of such new applications without the availability of monopoly rights. A producer-centered solution to this problem would be to grant pharmaceutical firms additional monopoly rights to new applications in such cases (Roin 2013). A free innovator-centered solution, in contrast, would be to support patients' and clinicians' capability to design and carry out clinical trials independent of producers. As was noted in chapter 10, the practicality of that route has been demonstrated in a trial of potential therapies for ALS (Wicks, Vaughan, Massagli, and Heywood 2011).

Extending Free Innovation Paradigm Insights beyond Innovative Content

Unpaid individuals in the household sector produce many socially valuable free information outputs in addition to innovation. Examples include collection, assessment, and diffusion of news by on-the-spot amateur observers (Benkler 2006), research and writing related to free contributions to Wikipedia, and the creation and free distribution of "fan fiction" by communities of unpaid amateur writers (Jenkins 2008; Jenkins, Ford, and Green 2013). These specific forms of non-innovative creative output from the household sector, and many others as well, are often collectively referred to as "user generated content" (UGC) or "user created content" (UCC). An OECD study defines "user generated content" as "i) content made publicly available over the Internet, ii) which reflects a 'certain amount of creative effort', and iii) which is 'created outside of professional routines and practices'" (Wunsch-Vincent and Vickery 2007, 4).

I propose that the activities and economic considerations involved in generating and diffusing UCC can be quite well described by the free innovation paradigm. After all, UCC, like free innovation, is generally developed by unpaid individuals motivated by self-reward and working in their discretionary time, and is generally not protected from free adoption by its developers.

Upon reflection, the usefulness of the free innovation paradigm for describing creative activities and outputs in the household sector beyond innovation will not be surprising. Many of the unique behaviors and the difficult policy choices associated with the producer innovation paradigm spring from producers' needs to capture monopoly profits from sales to gain private returns from their private investments in innovation. In contrast, innovation development within the free innovation paradigm is self-rewarded and therefore viable even if the outputs are diffused "for free." This also applies to non-innovative user-generated content produced with self-reward as a motivation and given away.

To illustrate the similarities, consider the writing and free distribution of "fan fiction" by unpaid, self-rewarded writers in the household sector. Writers of fan fiction generally base their works on the books of well-known authors. These "derivative works" are illegal under copyright law, but are nonetheless created and widely distributed for free by authors of fan fiction (Jenkins 2008). Individuals in the household sector who build upon the "platform" or "toolkit" inadvertently offered by copyrighted works create the same economic interaction effects with publishers that were discussed in chapters 6 and 7 with respect to interactions between free innovators and producers. Today, commercial publishers and popular authors are increasingly understanding that fan fiction is a commercially valuable free complement to their intellectual property, and so increasingly seek to support fan fiction rather than suppress it (Arai and Kinukawa 2014). Consumers of fan fiction prove to be avid buyers of the commercial works upon which fan fiction works are based. Indeed, fan fiction appears to expand the market for published fiction—to be a valuable free complement for producers. Further, just as designs created by free innovation are sometimes commercialized, fan fiction can be a source of commercially valuable writings and of new authors for commercial publishers (Jenkins, Ford, and Green 2013). In net, it appears that the economic interactions between the free fan fiction writers and commercial fiction producers are very similar to those described by Gambardella, Raasch, and von Hippel (2016) in the case of interactions between innovators operating within the free innovation and producer innovation paradigms.

Diffusion failures characteristic of the free innovation paradigm (described in chapter 5) can also affect user created content made available for free. For example, it has been found that many Wikipedia contributors, motivated by self-reward, choose to write on topics of personal interest rather than on topics of demonstrably stronger interest to larger numbers of Wikipedia readers. Thus, if a self-rewarded contributor of articles to Wikipedia is passionate about orchids, an article on orchids it will be—even if most Wikipedia readers would greatly prefer an additional article on plumbing. This pattern was confirmed by Warncke-Wang, Ranjan, Terveen, and Hecht (2015), who analyzed Wikipedia editions in four languages and found extensive misalignment between production and consumption in all of them.

I suggest that it will be very useful to explore how the principles of and practices within the free innovation paradigm can be extended beyond innovation to explain and support a wide range of personally and socially valuable development work in the household sector. Again, as many authors cited in this book make clear, free innovation in particular, and free creative activity in general, enhance both social welfare and many individuals' lives via such personally valued dimensions of experience such as self-expression and competence (Fisher 2010; Benkler 2006).

In this book I have sought to integrate new theory and new research findings, developed together with valued colleagues during the past few years, into the framework of a "free innovation paradigm." I have positioned the free innovation paradigm both as a challenge to the adequacy of the Schumpeterian innovation paradigm and as a useful complement. Both paradigms describe important innovation processes, with the free paradigm codifying important phenomena in the household sector that the producer innovation paradigm does not incorporate.

Recall that by proposing and describing the free innovation paradigm, I by no means claim that research needed to support it is complete. Indeed, I claim precisely the opposite. A new paradigm is most useful when understandings of newly observed phenomena are emergent and when ideas regarding a possible underlying unifying structure are needed to help guide the new research (Kuhn 1962). This is the role that I hope the free innovation paradigm described in this book will play. If it is successful, it will usefully frame and support important research questions and findings not encompassed by the existing Schumpeterian producer-centered paradigm, and so provide an improved platform for further advances in innovation research, policymaking, and practice.

The free innovation paradigm also, as a description of "democratized" household sector innovation practice, will help us expand our understanding of our personal freedoms and potential for creative action. By exploring more deeply what free innovation is and can become, we can more effectively support its growth and development—and thereby our own.

Appendix 1 | Household Sector Innovation Questionnaire

Cross-national comparisons of household sector activity in product development, such as those shown in chapter 2, were possible because our group of colleagues intentionally used the same basic questionnaire, sometimes with additional questions added, to conduct their surveys. Here, on the chance that additional colleagues might wish to extend the collection of comparable data to additional countries or within another country longitudinally, I reproduce the latest version of our joint questionnaire (de Jong 2016). Jeroen P. J. de Jong is the main author of this questionnaire. An expert on matters of questionnaire design and analysis, he offers to advise any who might wish to use or modify this questionnaire.

Please note that so far my colleagues and I have used the questionnaire reproduced below only to collect information on household sector *product* innovations, as in the national surveys discussed in chapter 2. In view of the economic importance of services in national economies, it would clearly be useful to extend data collection to household sector development in services too. Unfortunately, however, my colleagues and I have not yet found a reliable way to do this. We have tried several approaches and several variants of the questionnaire without success. The basic problem appears to us to be that respondents are generally unable to isolate instances of service development from non-innovative general patterns of day-to-day living when confronted with a questionnaire.

For example, respondents, assisted by cues, find it relatively easy to recall physically modifying a grandpa's favorite chair to make it easier for him to safely stand up from a sitting position—this would be a product innovation. On the other hand, the respondents appeared unable to recall and report having devised a sequence of special lifting movements to provide grandpa with the service of safely arising from his unmodified chair with caregiver assistance. This is the case even if follow-up, face-to-face interviewing finds that they had in fact devised such a sequence of lifting movements. The problem remained even when we start our questioning by asking respondents to recall a problem they had recently encountered, rather than a solution they had developed. They might then respond "I had a problem of getting grandpa safely up from his favorite chair." But when we next asked, "What did you do about it?," they appeared to us to be much more likely to recall making a physical modification to the chair than to recall making a service modification or a technique modification to solve the problem.

My colleagues and I do not think this reflects a general absence of service innovations by free innovators in the household sector. Recall that many of the respondents in the study of medical patient service innovations presented in chapter 8 were experiencing significant day-to-day difficulties and suffering from rare diseases. Contrary to our experience with general household innovators, who probably were generally experiencing and addressing less intense needs, these individuals often recalled service innovations. At least, they recalled those that provided them with significant help and improvements to their daily situation (Oliveira, Zejnilovic, Canhão, and von Hippel 2015). Researchers using the questionnaire reproduced below may wish to experiment with solving this problem—it will be important for all of us to have a good solution.

I next add a few additional comments to explain more about the choices we made in designing the questionnaire.

First, in order to aid respondents' ability to recall an innovation they may have developed, de Jong devised a procedure involving offering respondents a series of subject-specific cues—for example, "Did you within the last three years develop or modify computer software? household fixtures or furnishing?" The cues used are shown in the questionnaire reproduced below.

Second, note that the questionnaire includes screening questions to ensure that what is being described by a respondent as an innovation meets the criteria of the study. Respondents were asked whether they had created the innovation within the past three years, whether it was for their job or business (to screen out job-related innovations), and whether they could have bought a similar product on the market (to screen out homebuilt versions of existing products). Other screening questions could be added for other purposes.

Third, and again with respect to screening, we found it is useful to include an open-ended question asking respondents to briefly describe their claimed innovation. Such a question can help to exclude false positives. As experience shows, many householder respondents have only a vague idea of what the term innovation means. For example, one respondent said "Yes, I have innovated" and then proceeded to answer the screening questions listed above in a way that matched the study's criteria for an innovation. However, when asked to briefly describe the claimed innovation, the respondent stated "I built a new barn for my horses." This clearly false positive was caught only because of the inclusion of a request for a brief description of the innovation.

For more information on sample selection and more methodological information, see von Hippel, de Jong, and Flowers 2012, de Jong, von Hippel, Gault, Kuusisto, and Raasch 2016, and Kim 2015.

Survey Script

The following survey script is taken from de Jong (2016). Preceding this script, an introductory statement is provided to respondents, offering information on the purpose of the study and providing information on sponsorship, on how the data will be used, and on the confidentiality of answers. Section A is meant to identify respondent consumers who have innovated. Section B includes the main follow-up questions that have been used in empirical studies to date.

Section A

My next questions relate to any creative activities in your leisure time. You may have created novel products or product modifications for personal use, to help other people, to learn or just for fun. I will provide some examples.

A02. First, creating computer software by programming original code. Within the past three years, did you ever use your leisure time to create your own computer software?

1: yes 2: no

if A02>1 Go to A12

A03. Did you do this primarily for your employer or business?

1: yes 2: no

if A03 = 1 Go to A12

A04. At the time you developed it, could you have bought ready-made similar software on the market?

1: yes 2: no

if A04 = 1 Go to A12

A05. Did you primarily create it to sell, to use yourself, or for some other reason?

1: to sell 2: to use myself 3: other, please specify........

If A05 = 1 Go to A12

A06a. What kind of software did you create? [open answer]

A06b. What was new about this software? [open answer]

(Repeat the sequence of questions shown above for each of the following cues)

A12. The second example is household fixtures and furnishing, such as kitchen- and cookware, cleaning devices, lighting, furniture, and more. In the past three years, did you ever use your leisure time to create your own household fixtures or furnishing?

1: yes 2: no

A22. Next, you may have developed transport or vehicle-related products, such as cars, bicycles, scooters or anything related. In the past three years, did you ever use your leisure time to create your own transport or vehicle-related products or parts?

1: yes 2: no

A32. Tools and equipment, such as utensils, molds, gardening tools, mechanical or electrical devices, and so on. In the past three years, did you ever use your leisure time to create your own tools or equipment?

1: yes 2: no

A42. Sports-, hobby- and entertainment products, such as sports devices or games. In the past three years, did you ever use your leisure time to create your own sports-, hobby- or entertainment products?

    1: yes 2: no

A52. Children- and education-related products, such as toys and tutorials. In the past three years, did you ever use your leisure time to create your own children- or education-related products?

    1: yes 2: no

A62. Help-, care- or medical-related products. In the past three years, did you ever use your leisure time to create your own help-, care- or medical-related products?

    1: yes 2: no

A72. Finally, in the past three years, did you ever use your leisure time to create or modify any other types of products?

    1: yes 2: no

(follow-up questions and routing A13-A16b, A23-A26b, etc., see A03-A06b)

If number of valid innovations (A05, A15, … , A75 > 1) = 0 Go to End

If number of valid innovation = 1 Go to B01

A99. You just mentioned a number of creations. Which one do you consider most significant? 1: computer software 2: household or furnishing product 3: transport or vehicle-related product 4: tool or piece of equipment 5: sports-, hobby- or entertainment product 6: children- or education-related product 7: help-, care- or medical-related product 8: other product or application

Section B
My next questions are concerned with this specific [insert name of innovation that respondent identified in A99 as "most significant"] that you created. I will refer to it as the 'innovation'.

B01. Why did you develop this innovation? I will give you a list of reasons. Please indicate their importance by assigning zero to 100 points to each reason. The total number of points for all reasons together must add up to 100.

B01a: I personally needed it ____ points

B01b: I wanted to sell it/make money ____ points

B01c: I wanted to learn/develop my skills ____ points

B01d: I was helping other people ____ points

B01e: I did it for the fun of doing it ____ points

B02a. Did you work with other people to develop this innovation?

  1: yes 2: no
  If B02a = 2 Go to B03

B02b. How many others contributed to developing this innovation? ...persons

B03. Can you estimate how much time you invested developing this specific innovation? .... hours/days/weeks during ... days/weeks/months

B04a. Did you spend any money on this innovation?

  1: yes 2: no
  If B04a = 2 Go to B05

B04b. Can you estimate how much? ....Euros

B05. Did you use any methods to protect this innovation? (For example patents, trademarks, copyrights, confidentiality agreements)

  1: yes 2: no

B06. Supposing that other people would be interested, would you be willing to FREELY share what you know about your innovation?

  1: yes, with anyone 2: yes, but only selectively 3: no

B07. Supposing that other people would offer some kind of COMPENSATION, would you be willing to share your innovation?

  1: yes, with anyone 2: yes, but only selectively 3: no

B08. Did you do anything to inform other people or businesses about your innovation? (For example: Showing it off, communicating about it, posting its design on the Web)

  1: yes 2: no

B09a. To the best of your knowledge, have any other people adopted your innovation for personal use?

> 1: yes 2: no
> If B09a = 1 Go to B10a

B09b. Do you intend to contact other people who may adopt your innovation for personal use?

> 1: yes 2: no

B10a. Do you, alone or with others, currently own a business you help manage, or are you self-employed?

> 1: yes 2: no
> If B10a = 2 Go to B11a

B10b. Did you commercialize your innovation via your business? Or do you intend to do this?

> 1: yes, I commercialized it 2: yes, I intend to do so 3: no
> Go to B12

B11a. Are you currently, alone or with others, trying to start a new business?

> 1: yes 2: no
> If B11a = 2 Go to B12

B11b. Do you intend to commercialize your innovation with this start-up?

> 1: yes 2: no

B12a. Finally, commercial businesses like your employer or any other organization may be interested in your innovation. Did any commercial business adopt your innovation for general sale?

> 1: yes 2: no
> If B12a = 1 Go to End

B12b. Do you intend to contact commercial businesses to adopt your innovation for general sale?

> 1: yes 2: no

Appendix 2 | Modeling Free Innovation's Impacts on Markets and Welfare

In chapter 6, I summarized and discussed the findings of the modeling presented in Gambardella, Raasch, and von Hippel (2016). The model itself is significantly richer than a non-mathematical summary can convey, and so in this appendix I reproduce the original version of our model "set-up" information, the mathematical version of the model itself, and related findings exactly as presented in sections 4 and 5 and the appendixes of Gambardella, Raasch, and von Hippel (2016). Before reading this appendix, readers might wish to review chapter 6 above for contextual information not repeated below.

In my description of the model and the findings in chapter 6, I changed the term "user innovators" (used in Gambardella, Raasch, and von Hippel 2016) to "free innovators." However, I have kept the original term, "user innovator," in this appendix. I have done so because in the research article we defined user innovators as having exactly the same range of possible self-reward types as free innovators. The only difference is that in the article it was assumed that user innovators always reaped *some* level of self-reward from personal use, perhaps in addition to other possible types of self-rewards. This assumed pattern is likely to be very generally the case among real-world free innovators. Recall that some level of use motivation was present in all the free innovator cluster data shown in figure 2.1 of the present book.

Editor's note As was explained above, the remainder of this appendix is quoted from Gambardella, Raasch, and von Hippel (2016). The material is reproduced here as supplied by the author. The original numbering of sections and subsections has been retained. The figure, which was numbered 2 in the original paper, is referred to here simply as "the figure."

Section 4. Model set-up and findings

4.1 User types and 'tinkering surplus'

We divide a producer's potential market into two types of users: *innovating users* and *non-innovating users*. *Innovating users* find it viable to develop and self-provision innovative designs related to the producer product, e.g., improvements, customizations, and complements. They can also viably self-provision home-made copies of the producer product itself, and so can choose whether to buy the product from a firm or to make it themselves. *Non-innovating users* do not have a viable option of innovating. Their costs may be too high, for example, because they lack needed skills or access to tools, or because they have a high opportunity cost for their time. However, it is viable for non-innovating users to make copies and self-provision products based on designs developed by user innovators at some level of quality ranging from equal to innovating users down to zero.

The share of innovating users is σ, and we regard this share as exogenous and static; users cannot change their type. For simplicity, we normalize the size of the market to 1, so that σ and 1 – σ are also the number of users of each respective type.

With respect to the utility users derive from innovating, we note that empirical research finds that innovating users derive utility *both* from using the innovation they have created, and from innovation process benefits they gain from engaging in the innovation process itself, such as fun and learning (Lakhani and Wolf 2005; Franke and Schreier 2010; Raasch and von Hippel 2013). Users seek to maximize their utility from innovating, which we call h, by determining the optimal amount of resources, such as time t, to devote to innovation projects,

$$Max(t)h \equiv \chi + (\phi^{1-\alpha}/\alpha)x^{1-\alpha}t^{\alpha} + 1 - t. \tag{1}$$

In equation (1), the parameter χ represents a user innovator's utility, net of all innovation-related costs, from go-it-alone innovation projects, i.e., when producers do nothing to support him. The second term of (1) represents the user innovator's additional utility when a firm conducts x projects to support his endeavors. Examples of such support are the development of design tools for the use of users, and gamification to make product design activities more enjoyable to users. The parameter $\alpha \in (0,1)$ captures whether innovating users' utility is mostly determined by the time they invest (high α) or by the extent of firm support (low α). The parameter $\phi > 0$ captures the productivity of this process. The last term, $1 - t$, captures the value of the user innovator's remaining time that he can spend on other matters, when the total time he has available is normalized to 1 and he has decided to spend t on innovation projects.

We derive from (1) that the user's utility-maximizing time investment in innovation is $t = \phi x$, which yields utility $h = \chi + (1 - \alpha/\alpha)\phi x + 1$. We call this expression capturing users' net benefit from innovating the *tinkering surplus* (TS), where TS is the aggregate net benefit that all users gain from innovating and self-provisioning. It consists of benefits from the use of the self-provisioned innovation, plus innovation process benefits, as mentioned above, minus costs. When the investment of firms in user innovation support is zero, innovating users still get their go-it-alone tinkering surplus, $h = \chi + 1 > 0$. If firms do invest ($x > 0$), TS increases as a function of the level of that investment.

4.2 Shared vs. producer-only innovation

We decompose the value that all buyers derive from the producer product into two parts: value v that they derive from features and components that only the producer firm will develop and produce, and value b that buyers derive from features and components that can be developed and produced by firms *and* users, jointly or in isolation.

Features that only producers will find viable to develop include those that offer limited value to many individual users. No individual user would find it viable to develop such a feature, but producers can aggregate demand across buyers and thereby recoup their investment (Baldwin and von Hippel 2011). Features in this category may include, e.g., product engineering for greater durability and ease of use, a more elaborate design, a manual to accompany the product, etc. In contrast, features b that both individual users (typically "lead users") and producers can viably develop require smaller investments, compensated for by larger benefits to individual user innovators. They provide high functional novelty and solve important, hitherto unmet user needs (von Hippel 2005). As the needs of lead users foreshadow demand in the market at large (cf. definition of lead user), non-innovating users too will predictably benefit from solutions to these problems with the passage of time.

We assume that all users tend to have more similar assessments of the features we call b, that innovative users may get involved in developing, than of the features v that the producer has to develop on its own. Capturing this idea of less heterogeneity with regard to b but simplifying our analysis, we assume that users differ only in their valuations of v ($v \sim U[0, 1]$) whereas they all like b to the same degree. In our model of innovation and production by users and producers, we focus on innovations of type b, following our assumption that producers are the only ones to invest in v. Innovations with regard to b are assumed to depend on two activities.

First, the volume of innovations of type b depends on the aggregate effort T exerted by all innovating users, to the extent that it is useful to the firm (e.g., net of redundancy). To streamline our analysis, we assume that the aggregate usable effort is simply proportional to the total efforts t of the σ innovating users, that is $T = \gamma' \sigma t$, $\gamma' > 0$. (We could use more complex aggregations, allowing for increasing or diminishing returns to the number of innovating users, but our results would remain materially unchanged.) Assuming identical innovating users, and employing the optimal expression for t, $t = \phi x$, we obtain aggregate user effort

$$T = \gamma \sigma x$$

where $\gamma = \gamma' \phi$ comprises any factor that raises the ability of the firms to take advantage of the productivity of the innovating users' efforts to improve b. As explained earlier, the firm can influence aggregate user effort T through x projects to develop tools and platforms that support and leverage innovating users. The projects affect the time t users want to spend on innovation projects, which then affects the value of the innovative product b via aggregate effort T.

Second, innovations of type b are a function of some commitment of resources Y carried out by the firm. To fix ideas, Y can be commercial

R&D projects or any other product creation or development activity. We define

$$Y = \xi(1-s)y, \quad \xi \geq 0$$

where y is the total number of innovation projects of the firm. The firm allocates a share s to projects that support innovating users, that is $x = sy$, and the remainder, $(1-s)y$, goes to traditional commercial R&D projects (either in-house or external). Projects that support innovating users are of little commercial value, per se, but indirectly produce value by attracting more user innovation activities. The parameter ξ measures the productivity of the firm's commercial R&D.

Taking into account these two drivers of innovation—aggregate user effort T and producer R&D activity Y—let the value of the innovative product to users be

$$b = (T^{\beta} + Y^{\beta})^{1/\beta}, \quad \beta > 0,$$

which we can rewrite as

$$b = \left[\tau^{\beta} s^{\beta} + \xi^{\beta} (1-s)^{\beta} \right]^{1/\beta} y = \tilde{b}y$$

where $\tau \equiv \gamma \, \sigma$ and $\tilde{b}\left[\tau^{\beta} s^{\beta} + \xi^{\beta} (1-s)^{\beta} \right]^{1/\beta}$ is the productivity of all the firm's y projects taken together.

4.3 User and producer innovation activity as substitutes or complements

The parameter β plays an important role in our analysis. It captures two options that firms can choose from, each of which involves a distinct form of organizing tasks and resources for innovation. The first option is such that the efforts of innovating users, T, and those of the producer, Y, are substitutes. Take, for instance, the writing of new software code. Suppose that both the producer and users can work on each of two tasks, (1) novel functionality, and (2) the creation of convenience-enhancing features such as "user friendly" installation scripts. The more effort the producer spends on each of these tasks, the lower the innovation impact that users can make, and vice versa. One effort tends to substitute for the other. In our model, this situation is captured by $\beta > 1$, which implies that the marginal impact of T on b decreases as Y increases and vice versa.

The second option, in contrast, structures R&D for complementarity between user and producer innovation activities. In our example, suppose that users write novel code and producers develop "convenience features." The more effort users put into coding, the higher the impact that producers can make, and vice versa. In our model, this situation is described by $0 < \beta < 1$, which implies that the marginal impact of T on b increases as Y increases and vice versa. Research has shown that user innovators tend to focus on developing innovations providing novel functionality, and producers on developing innovations that increase product reliability and user convenience (Riggs and von Hippel 1994; Ogawa 1998). A good example in the software field is RedHat. That firm's commercial offerings are based on open source software code such as Linux and Apache software, developed by users, to which RedHat adds convenience features such as "easy installation" software scripts.

To streamline our analysis, we assume that each firm can pick its preferred innovation option, but not the specific level of β. A fully endogenous β would add complexity without substantial new insight. In practice, its value will depend on the industry in question, the technologies available to the firm, and best practices for integrating innovating users in R&D.

4.4 Individual market demands of innovating users and non-innovating users

Next, we need to understand the demand for the producer product from non-innovating users and from innovating users given user contestability, user-created complements and spillovers, i.e., the different types of interactions that we developed in section 3 [of this paper].

Starting with innovating users, we expect that they will buy the product from a firm only if their consumer surplus is positive and exceeds their surplus from self-provisioning, i.e., if

$$v + b - p + h \geq \lambda b + h, \, v \sim U[0, 1], \qquad 0 \leq \lambda \leq 1. \tag{2}$$

The term $v + b - p$ is the consumer surplus, where $v + b$ is our value decomposition of the producer product (cf. section 4.2) and p is its price. In case of self-provisioning, a user innovator will not get utility v, which is provided by the firm only. Of utility b that all innovating users co-create with the firm, he will get only the "walk-away value" λb that he can realize by learning from this co-creation process and trying to build features akin to b on his own. The quality $0 \leq \lambda \leq 1$ of his self-provisioned version of b will depend on several factors, such as the extent and format of information spillovers from the firm to the user innovator, his "absorptive capacity" for the spillovers, and his skills to build the information into a usable artifact. In the case of software programming, for instance, where the producer opens up his source code for users to co-develop, λ will be close to 1, if and as the essential design information required to replicate functionality b is fully revealed. In this example, if the producer shares only part of his source code, λ is depressed accordingly.

Finally, recall the user innovator's surplus h from her own innovation activities, including those extensions and customizations that the firm is not interested in. The user innovator is assumed to get this surplus h—the tinkering surplus—regardless of whether she buys the producer product or not.

Recall that non-innovating users simply buy a producer-provisioned product via the market or, to the extent that they are able, can elect to replicate a design developed and then shared peer to peer by a user innovator. Building on what we said earlier about v, b and p as constituents of demand, we expect that non-innovating users will buy on the market if

$$v + b - p + \mu'h \geq \mu b + \mu'h, \qquad 0 \leq \mu, \qquad \mu' \leq 1 \tag{3}$$

and self-provision otherwise.

The parameters μ and μ' in equation (3) capture the non-innovating users' ability to obtain knowledge of the innovating users' designs (which will depend on the innovating users' propensity to diffuse design information), to replicate them, and to benefit from them. While μ' refers to a non-innovating user's ability to benefit from an individual user innovator whose design they adopt, μ captures their trickle-down benefits from what the user innovator has learned from the producer as well as other innovating users during the co-creation process of b. Of course, when the non-innovating users buy from the firms they enjoy b incorporated in the firms' product, while they enjoy μb when they obtain the product from the innovating users through peer-to-peer diffusion. We expect non-innovating users to have imperfect knowledge of the innovating users' designs, to be less skilled at self-provisioning them and/or to benefit less from using them ($\mu \leq \lambda$ and $\mu' \leq 1$). With respect to imperfect knowledge and higher costs of self-provisioning, consider that innovating users may well regard careful design documentation for the benefit of potential adopters to be an unprofitable chore in the case of freely revealed designs (de Jong et al. 2015; von Hippel, DeMonaco, and de Jong 2016). With respect to lower levels of benefit, consider that the designs were developed to precisely suit the innovating users' individual tastes.

Finally, it is crucial to note the trade-off that our model implies for the producer: Firms benefit from learning from innovating users about how to make a better product b for both innovating and non-innovating customers; to that end they want to invest in x to involve users more extensively. At the same time, this comes at the cost of facilitating self-provisioning by both innovating and non-innovating users. As the producer invests in tools and toolkits, modularizes the product or reveals design knowledge such as source code to facilitate user innovation, he also makes it easier for both innovating and non-innovating users to self-provision rather than buy. Our model assumes that the producer cannot entirely avoid this side effect of enhanced user contestability, even while choosing a mode of supporting user innovation that best serves his goals.

4.5 Profit maximization by firms

The aggregate demanded quantity of $(1 - \sigma)$ non-innovating users and σ innovating users is

$$q = (1 - \sigma)(1 - p + (1 - \mu)b) + \sigma(1 - p + (1 - \lambda)b) = 1 - p + \eta b, \qquad (4)$$

with

$$\eta \equiv (1 - \mu)(1 - \sigma) + (1 - \lambda)\sigma.$$

Solving for p, inverse demand is

$$p = 1 + \eta b - q. \qquad (5)$$

With N symmetric firms in the market, aggregate demand is $q = \sum_{j=1}^{N} q_j$ and q/N is the demand faced by one firm. Firm profits Π_i are given by the number of units sold by firm i, q_i, times the profit margin, given by price p minus marginal cost of production φ, and minus the cost of y innovation projects,

$$\Pi_i = (p - \varphi)q_i - \kappa y^2, \qquad \kappa > 0 \qquad (6)$$

where we assume diminishing returns to running y projects.

In order to maximize profits, firms make several interrelated decisions in the following sequence: First, they decide on the organization of their R&D. Specifically, they pick one of two options available to them: the organization of R&D such that user and producer inputs, T and Y, are substitutes ($\beta > 1$) or the organization for complementarity ($0 < \beta < 1$). It will take firms longer to change their organizational structure and capabilities in R&D than to change the number of projects, which is why we model this is as the first choice. Next, the firms pick their total number of R&D-related projects (y). Then they decide on the share of projects ($1 - s$) to allocate to traditional producer R&D. The remainder of the projects, share s, will be devoted to user-innovation support and thus indirectly increase the flow of new product ideas available to the firm. Finally, firms decide on the quantity to produce and sell on the market (q_i).

We use backward induction to derive the producers' optimal decisions. In this section, we look at the optimal choices of q_i, s, and y, in this order. In section 4.7, we will study the choice of innovation mode (β).

Choice of q_i. We take the derivative of (6) with regard to output quantity (q_i) and obtain *foc*: $1 + \eta b - \varphi - \sum_{j=1}^{N} q_j - q_i = 0$. In symmetric equilibrium, this produces profit-maximizing quantity, price and profits

$$q_i = (1 + \eta b - \varphi)/(N + 1) \qquad (7a)$$

$$p = (1 + \eta b - \varphi)/(N + 1) + \varphi \qquad (7b)$$

$$\Pi_i = (p - \varphi)^2 - \kappa y^2 = [(1 + \eta b - \varphi)/(N + 1)]^2 - \kappa y^2. \qquad (7c)$$

Choice of s. To determine the share of the firms' projects aimed at supporting user innovation, s, we maximize $\tilde{b}^U = (\xi^\theta + \tau^\theta)^{1/\theta}$ yielding *foc* $\tau^\beta \beta s^{\beta-1} - \beta \xi^\beta (1-s)^{\beta-1} = 0$. To determine the optimal s, a case distinction is required. In the case of complementarity between user efforts and producer R&D, i.e., if $0 < \beta < 1$, the *soc* is negative, which implies that there is an intermediate project allocation $0 < s < 1$ to user support that maximizes innovation output $\tilde{b}$ (specifically $s = \tau^\theta/(\xi^\theta + \tau^\theta)$, with $\theta \equiv \beta/(1-\beta)$). As can be seen from the expression for τ, this optimal project share allocated to user innovation supports increases in the share of innovating users in the market and their productivity in terms of commercially valuable ideas ($s_\sigma, s_\gamma > 0$, where from now on we use subscripts to denote derivatives), and decreases with the productivity of producer R&D ($s_\xi < 0$). In the case of substitution between user and producer innovation efforts, i.e., if $\beta > 1$, the *soc* is positive, which implies that the optimal allocation to user support, s, is either 0 or 1, depending on whether the productivity of the user contribution in $\tilde{b}$, that is τ, is greater or smaller than the productivity of the firm contribution, ξ.

Choice of y. The *foc* of (7c) with respect to y is $2(1 + \eta b - \varphi)\eta \tilde{b} / (N+1)^2 - 2\kappa y = 0$, which yields $y = (1 - \varphi)z/[\kappa(N+1)^2 - z^2)]$, where $z \equiv \eta \tilde{b}$. Note that the *soc* implies $\kappa(N+1)^2 - z^2 > 0$ such that the profit-maximizing investment y is always positive. It is also easy to see that y increases with z.

4.6 The producer vs. user-augmented innovation modes

From our findings from the previous section relating to the distribution of innovation projects by the firm (s), we see that there are two modes of innovating, and that firms will want to choose between them. The first mode is characterized by $\beta > 1$ and $s = 0$. That is, in this mode firms choose to organize their R&D such that user and producer efforts are substitute inputs and then allocate their entire budget to their own commercial R&D efforts, not supporting user innovation activities in any way. We call this the producer (P) *innovation mode*. In this mode, firms ignore the innovating users and organize the creation of b solely around closed commercial R&D.

As a consequence of being closed, firms need not fear information spillovers to innovating users ($\lambda = 0$) and on to non-innovating users (μ

= 0). In the producer mode, therefore, the demands of the non-innovating users and the innovating users simplify to

$$v - p + b + \mu'h \geq \mu'h \tag{8}$$

$$v - p + b + h \geq h, \tag{8'}$$

respectively. At the same time, aggregate demand is (4), with $\eta = 1$ rather than $(1 - \mu)(1-\sigma) + (1 - \lambda)\sigma$.

The second innovation mode is characterized by the firm organizing its R&D for complementarity with user innovators ($0 < \beta < 1$) and then making a positive investment in user innovation support (optimal $s = \tau^\theta/(\xi^\theta + \tau^\theta) > 0$). We call this the *user-augmented* (U) *mode*. In this mode, firms actively leverage user-created spillovers for innovation and organize their R&D to exploit the complementarity between the two sources of innovation. Users contribute to raising the use value b of the product, which enhances the demand of both the non-innovating users and the innovating users. At the same time, firms' support of innovating users creates user contestability with regard to features b ($\lambda,\mu \geq 0$).

To summarize, the trade-off between the U- vs. P-modes pivots on producers investing to facilitate user innovation and reap spillovers, but by this action simultaneously and unavoidably boosting user self-provisioning to a degree that may be small or large.

4.7 Choice of innovation mode (β)
Continuing our earlier process of backward induction to understand outcomes in markets with innovating users, we now consider the very first producer decision, the choice of innovation mode. Our goal is to understand under what conditions a producer will prefer the producer mode over the user-augmented mode, or *vice versa*. Additionally and importantly, we examine under what conditions the increasing prevalence of innovating users that we observe in many markets (cf. Baldwin and von Hippel 2011) renders user integration the profit-maximizing innovation strategy for producers.

Our first theorem below explains the choice of innovation mode by a producer firm. It establishes that, subject to two conditions, firms in markets with an increasing share of innovating users will find it in their own best interest to switch to the user-augmented mode. In switching, firms are aware that they are strengthening user contestability, but also realize that, overall, this is more profitable than a closed innovation approach.

To find the profit-maximizing mode of innovation, it is convenient to rewrite expression (7c) for the profits of the firms as

$$\Pi = [(1 + zy - \varphi)/(N + 1)]^2 - \kappa y^2. \tag{9}$$

This expression captures profits in both the P- and U-modes, which differ only in z. (In particular, in the P-mode $z^P = \eta^P \tilde{b}^P$, with $\eta^P = 1$ and $\tilde{b}^P = \xi$; in the U-mode $z^U = \eta^U \tilde{b}^U$, with $\eta^U = (1 - \mu)(1 - \sigma) + (1 - \lambda)\sigma$ and $\tilde{b}^U = (\xi^\theta + \tau^\theta)^{1/\theta}$.) Given the optimal choices of s and y, as derived in section 4.5, this implies that $\Pi^P \geq \Pi^U$ if and only if $z^P \geq z^U$. In other words, we can check whether profits are higher in the P- or U-mode simply by checking whether z is higher in one or the other.

We find that, when there are very few innovating users (σ close to zero), profits in the P-mode are always higher than profits in the U-mode ($\Pi^P > \Pi^U$). Thus, when there are very few innovating users, firms choose the P-mode. The intuition is that, from the firms' perspective, the user innovation spillovers that they can harvest, the upside of conducting projects x to support user innovation, are low. At the same time, the downside is considerable, as the information and tools that the firms supply to the few innovating users can enable innovating users to develop a competing design and share it peer to peer, knocking off a good part of the producer's demand. The magnitude of this loss, and thus the downside of switching to the U-mode, will depend on λ and μ, users' ability to self-provision b.

As the share of innovating users increases, profits stay the same in the P-mode but increase in the U-mode. (This is true under two conditions that we will explain below.) Firms will switch from the P- to the U-mode when the share of innovating users is larger than a threshold σ^*, beyond which $\Pi^U > \Pi^P$. This is illustrated in [the figure].

The first condition relates to λ and μ. When user contestability is very weak (as indicated by the uppermost curve for which $\lambda = \mu = 0$), the producer can switch to the user-augmented mode free of risk. On this curve, when the share of innovating users is $\sigma = 0$, profits are equal for both modes of innovating. Then, as σ increases, the U-mode outpaces the P-mode in terms of firm profits. Intuitively, in this case firms

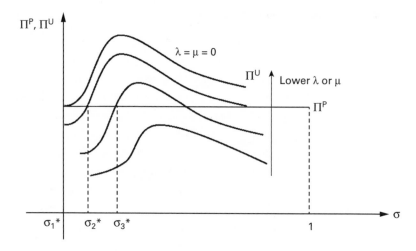

Firms' profits under the U- and P-modes.

benefit from the contribution of innovating users without risking the rise of self-provisioning and concomitant reduction of demand for the firms' product. When user contestability is more pronounced (as illustrated by the second and third curve), we see that the threshold σ* at which the switch to the U-mode can occur shifts to the right—that is, a higher share of innovating users in the market is needed for the producer to prefer the U-mode. If spillovers λ or μ are very large, as illustrated by the bottom curve, a switch to the U-mode will never be attractive to firms.

The second necessary condition for the switch to occur is that the complementarity between user and producer efforts T and Y must be strong enough. Specifically, $\theta < 1$ (i.e., $0 < \beta < \frac{1}{2}$) must hold.[1] In other words, the contribution of the innovating users must be strong enough to trigger a significant increase in b that outweighs the negative impact on profits from intensified user contestability; otherwise firms will prefer to stay in the producer mode.

Theorem (Choice of mode). If the innovative contribution of the innovating users is sizable ($0 < \beta < \frac{1}{2}$) and user-contestability (λ and μ) is not too high, a critical mass of innovating users ($\sigma > \sigma^*$) makes profit-maximizing firms prefer the user-augmented mode of innovating to the producer mode.

Proof. See Appendix A [of this paper].

It should be noted that while firms may find it profitable to switch to the U-mode at threshold σ^*, they may switch back again to the P-mode at a high σ. (As illustrated in [the figure], Π^U reaches a maximum and then declines, potentially even falling below the Π^P line.) This is particularly likely at higher levels of λ and μ (cf. [the figure]). The reason is the following: By our assumption of $\lambda \geq \mu$ innovating users are more capable than non-innovating users of self-provisioning, i.e., they exhibit a superior outside option and thus lower demand for the product of the firm. When the share of innovating users σ gets quite large, this not only means extensive user innovation spillovers to firms but also implies that the share of non-innovating users—those who benefit the most from these spillovers by getting to buy a superior product—is small. Having many innovating users implies having low demand, particularly if λ is large. This detracts from the attractiveness of the U-mode and may make firms prefer to switch back to the P-mode where they can better capture demand. We will leave this issue for future research to investigate in more detail, since our core objective is to understand the initial switch from the producer mode to the user-augmented mode when the prevalence of innovating users increases.

4.8 Welfare and policy

In this final section, we consider the welfare implications of firms choosing either to "go it alone" in the producer mode of innovating or to integrate user inputs in the user-augmented mode. We need to understand whether firms' choice of mode is efficient from a societal perspective and if not, whether policy is likely to improve economic outcomes.

Calculations of social welfare that include user innovation are different from the standard mode of calculating welfare. Conventionally, social welfare is calculated as profits (PS) plus consumer surplus (CS). When innovating users develop and build a new product for their own use, welfare calculations must be modified to include their full costs and benefits. In particular, we need to take into account their tinkering surplus TS, which is the aggregate net benefit that all users gain from self-provisioning, if they choose to do so. To give an example, if a user self-provisions a newly designed product at a cost of 10 dollars and receives a monetized use value of 30 dollars, her tinkering surplus equals 20 dollars. Recall that benefits to tinkering can also accrue in the form of process value (Franke and Schreier 2010; Raasch and von Hippel 2013), e.g., enjoyment of or learning from the innovation process itself, or social status in the user community. Our model is agnostic to the composition of these benefits. It only presumes them not to be profit based, in line with the definition of a user innovator. We will consider generalizations of this aspect in the discussion section.

Incorporating these considerations, then, welfare in markets containing both user and producer innovators should be computed as

$$W = PS + CS + TS \tag{10}$$

where PS and CS are the standard producer and consumer surplus and TS is the tinkering surplus. How significant is the omission of the tinkering surplus in conventional analyses? The answer depends on the extent of user self-provisioning in a market. If many users self-provision (as is common across an increasing range of markets, especially markets for digital products, cf. Baldwin and von Hippel 2011), the omission can be substantial. In some cases, it may dwarf traditional components of welfare.

In our model, the tinkering surplus for a user innovator equals h, while for non-innovating users it is $\mu'h$, which stems from their ability to tap into peer-to-peer diffusion from the innovating users. Computing the components of welfare as they accrue to producers (aggregate profits, PS), non-innovating users (consumer surplus, CS^{nui} plus tinkering surplus TS^{nui}) and innovating users (CS^{ui} plus tinkering surplus, TS^{ui}), we have

$$PS = N\Pi$$

$$CS^{nui} + TS^{nui} = (1 - \sigma) \left[(1 - p + (1 - \mu)b)^2/2 + \mu b + \mu'h\right]$$

$$CS^{ui} + TS^{ui} = \sigma \left[(1-p + (1 - \lambda)b)^2/2 + \lambda b + h\right]$$

The first term is the aggregate profit of all the producers. The second term is the aggregate surplus of all non-innovating users, calculated from

$$(1-\sigma)\left[\int_{p-(1-\mu)b}^{1} (v - p + b + \mu'h)dv + \int_{0}^{p-(1-\mu)b} \mu b + \mu'h \, dv\right].$$

The third term derives from

$$\sigma\left[\int_{p-(1-\lambda)b}^{1}[v-p+b+h]dv+\int_{0}^{p-(1-\lambda)b}(\lambda b+h)dv\right].$$

These expressions will differ, depending on whether the U-mode or the P-mode is being chosen by firms.

Our analysis of welfare produces two main results that we summarize in two theorems. The first theorem states that, given our condition $0 < \beta < \frac{1}{2}$, higher firm profits in the U-mode imply higher welfare in the U-mode, but the reverse is not true. That is, whenever firms' profits are higher in the U-mode, welfare is aligned; in contrast, when firms' profits are higher in the P-mode, welfare may not be aligned. Specifically, there are levels of σ, the share of innovating users in a market, such that profits are higher in the P-mode but welfare is higher in the U-mode. As a result, to the extent that the decision to switch belongs to the producers, as we modeled it, producers will remain in the P-mode even though the share of innovating users is substantial and social welfare would be better served in the U-mode. The reason is that firms do not internalize the key externalities of our model—that is, the increase in tinkering surplus (h) accruing to users because of firms' investment in user support (x) and also facilitation of self-provisioning that firms bestow on innovating users (λb) and, subsequently, non-innovating users (μb) even if they do not buy the product.

Theorem (Welfare). Under the conditions of the choice-of-mode theorem, if firms' profits are higher in the user-augmented mode, so is welfare, but the reverse is not true.

Proof. See Appendix B [of this paper].

Our second result regards policy. We show that policies that increase the productivity of innovating users *can never reduce welfare*, provided that the costs of such policies do not outweigh their benefits. By contrast, policies that increase the productivity of R&D within firms *may reduce welfare*.

Examples of policies that raise the productivity of innovating users, γ, are subsidized access to design tools and maker-spaces. If innovating users become more productive, both profits and welfare rise under the U-mode but not under the P-mode, which, after all, does not leverage users' productivity. As firms' profits in the U-mode increase, they may come to exceed profits in the P-mode. We know from the previous theorem that if profits are higher in the U-mode, welfare is also higher.

Policies that increase the productivity of producer R&D, ξ, include R&D subsidies and tax exemptions as well as publicly funded applied R&D. Increases in firms' research productivity ξ raise profits in both the P- and the U-modes. We show that, unless complementarity between user and producer efforts is high, increases in ξ induce a larger increase in profits in the P-mode than in the U-mode.[2] This means that policies that support traditional producer R&D may induce a switch back to the P-mode. Since welfare is sometimes lower in the P-mode even while firms prefer it, increases in ξ may render the P-mode more attractive to the firms in spite of the fact that welfare is higher in the U-mode. In other words, such increases may induce a switch to the P-mode while welfare is higher in the U-mode, or prevent a welfare-increasing switch to the U-mode.

To summarize, policies that support producer innovation productivity ξ may reduce welfare. The mechanism is that such policies encourage firms to adopt a closed producer innovation mode while welfare may be higher in an open user-augmented mode. In contrast, policies that support the productivity of the innovating users can never reduce welfare. This is because they can only encourage a switch to the U-mode and this is never welfare reducing because whenever firms prefer the U-mode welfare is higher in this mode.

Theorem (Policy). Under the conditions of the choice-of-mode theorem, policies that raise the productivity of innovating users, γ, encourage firms to adopt the user-augmented mode and can never reduce welfare. By contrast, if the complementarity between user and producer innovation activities, T and Y, is weak ($\beta > \beta^*$, with $\beta^* < \frac{1}{2}$), policies that raise firms' research productivity, ξ, encourage firms to adopt the producer innovation mode, which may reduce welfare.

Proof. See Appendix C [of this paper].

Section 5 and 5.1 Discussion

In this paper we analyzed the effects of user innovation by consumers on standard outcomes in markets for innovation. Our special focus was on understanding the implications of the increasing prevalence of innovating users (σ increasing from a low level), as found in many markets.

Our principal findings were three. First, as the share of innovating users in a market increases beyond a certain threshold, firms' profit-maximizing strategy is to switch from the traditional producer-only innovation approach to an innovation mode that harnesses user innovators. Subject to two intuitive conditions relating to the innovative and competitive impact of user activities, welfare is higher in this user-augmented mode than in traditional producer-only innovation mode. All of the constituencies—producers, innovating users, and non-innovating users—benefit.

Second, any firm that elects to switch to integrating innovating users definitely augments social welfare; but firms generally switch "too late." Thus, markets containing both user and producer innovators tend to fall short of their theoretical optimum in terms of value creation because producers are too slow, from a social welfare perspective, to embrace user innovation. Thus, producers' optimal R&D strategies yield a suboptimal division of innovative labor between users and producers at the societal level. Underlying this inefficiency there are externalities that the producer cannot capture, e.g., the "tinkering surplus" that accrues to users, a novel component of social welfare.

Third, policies that raise the productivity of innovating users encourage firms to switch to the user-augmented mode and can never reduce welfare. By contrast, policies that raise firms' research productivity encourage firms to switch back to the traditional producer-innovation mode and thereby may reduce welfare.

5.1 Assumptions, robustness and generalizability of findings

Our model rests on several assumptions that can be usefully investigated via further research.

First, as we mentioned at the start of the paper, innovating users are defined as individuals or firms developing innovations to use rather than sell. In this paper, we have focused on individual consumer innovators only. We have done this to highlight the contestable nature of their demand, and to emphasize that contestability can occur in markets for consumer goods. However, follow-on research could develop a similar model focused on or including user firms creating, for example, process innovations for their own use rather than for sale.

Second, we note that there are fields and markets in which some types of innovations originate only from innovating users—a situation with $s = 1$ in terms of our model. This is often the case, for example, with respect to the development of specialized techniques. Producers often find it impossible to profitably develop and market unprotectable techniques, and tend to leave that vital arena entirely or almost entirely to users (Hienerth 2016). In this paper we explored the importance of user innovation in markets that include producer innovation as well. However, further work could explore the nature of markets characterized by user innovation only.

Third, for simplicity, our model assumed that all innovating users will be able to benefit from a producer's investment in user innovation support, and that the producer will be able to observe the efforts of all innovating users and be able to reap any valuable spillovers. This is clearly not the case in practice—users will be differentially affected, and producers will not be able to observe or capture all spillovers generated by users. However, the same modeling logic and the same findings apply if our assumptions are true only for a subset of users.

Fourth, we assume that producers can choose the level of investment in support of innovating users that will maximize their profits. In the real world, users are independent actors who often have power to "push back" against producer plans and actions. They also can initiate user innovation activities in ways that producers do not expect. An example of investment in supporting user innovation not going according to producers' profit-maximizing plans is the case of Xara, a proprietary software company. In 2006, Xara invested in opening a large percentage of the source code of Xara Xtreme, a vector graphics package, as a way to invite user innovation. However, Xara did *not* open a small, commercially critical part of the source code. This omission caused a boycott among user programmers and, in the end, Xara yielded and opened more of the code than they would have preferred absent pressure from innovating users (Willis 2007).

It would be valuable and interesting for follow-on research to address situations such as the above. While in this paper we assumed that producers decide unilaterally to what extent they want to support and complement user innovation activities, we could think of a game in which innovating users can possess the power to determine the extent of user support, s, and potentially even the degree of complementarity, β. We expect that, in such a game, when the power to make both decisions lies with innovating users, they will pick higher levels of user support and complementarity than producers would. Unless users pick very high levels of s, this should lower producer profit but increase welfare overall. Future research could further explore this and also consider situations in which the decision power with regard to s and β is distributed between innovating users and producers.

Fifth, it is noteworthy that user innovators in our model receive no remuneration from producer firms. In the real world, successful user innovators sometimes receive payments for valuable contributions (such is the case with Lego and many app stores). Still, as a nationally representative survey in Finland shows, innovating users typically do freely reveal their innovations; our assumption of no payment is based on that situation (de Jong et al. 2015). In a different model, our variable x could be seen as the cost of user royalties to the firm, and implications for market outcomes could be explored.

Sixth, we have modeled producer support of user innovation as *increasing* the amount of time (or resources more generally) that users wish to spend on activities that benefit producers. Gamification of contributions and the setting-up of a user community were examples in point. It is also conceivable, however, that producer support, e.g., in the form of better tools, will enable users to *save time* while innovating. Such kinds of producer support could attract additional users to contribute, i.e., those who were previously non-innovators. This would endogenize the share of user innovators, σ, in a market, which we have taken to be exogenous in our model. It would be interesting for future research to explore the outcomes of this extended model, especially with regard to the optimal choice of producer strategy β.

Finally, our model treated all producers symmetrically, having all of them choose either a user-substituting or a user-complementing innovation strategy. Future research can usefully generalize from this limiting assumption. In the real world, we observe the coexistence of producers of both types. A key reason, we think, is that reorganizing and re-structuring R&D to exploit user-created innovation spillovers can be quite costly. Established firms with a legacy of producer-centric innovation will therefore be hesitant to switch, while new entrants without a commitment to the traditional model will likely find it economically more viable to choose the user-augmented innovation mode. Such constraints and switching costs could usefully be analyzed regarding their effects on strategic heterogeneity and firm and market-level outcomes. For instance, in markets with a growing share of user innovators, we should observe that new entrants and incumbents that are more flexible in organizing their R&D are more profitable.

Appendix A: Proof of the Choice-of-Mode Theorem

As noted, $z^U \leq z^P \rightarrow \Pi^P \geq \Pi^U$ and vice versa. Moreover, σ affects Π only through z and therefore we can study the impact of σ on Π by studying the impact of σ on z. We first show that at $\sigma = 0$, $z^U \leq z^P$, which establishes that at $\sigma = 0$ firms choose the P-mode. We then show that and if $0 < \beta < \frac{1}{2}$, $z_\sigma^U \geq z_\sigma^P$, $\forall \sigma < \sigma_0$, with $\sigma_0 < 1$. For the first point, compare $\eta^U \tilde{b}^U$ and $\eta^P \tilde{b}^P = \xi$. At $\sigma = 0$, $\eta^U \leq \eta^P$ and $\tilde{b}^U = \tilde{b}^P = \xi$. As a result, $\sigma = 0$ implies $z^U \leq z^P$. For the second point, it is not difficult to see that $z_\sigma^P = 0$, and $z_\sigma^U = \tilde{b}^U(\eta_\sigma^U + \eta^U \tilde{b}_\sigma^U / \tilde{b}^U)$. Recall that $\eta_\sigma^U \leq 0$, and it is easy to see that $\tilde{b}_\sigma^U / \tilde{b}^U = \tau^\theta / [\sigma(\xi^\theta + \tau^\theta)]$. This expression is positive, and if $\theta < 1$, or $0 < \beta < \frac{1}{2}$, it is very high when $\sigma \rightarrow 0$. Moreover, it declines as σ increases, and, given $\lambda \geq \mu$, η^U declines. All this implies that when σ is close to zero, $z_\sigma^U > 0$ because the positive value of $\eta^U \tilde{b}_\sigma^U / \tilde{b}^U$ outweighs the negative η_σ^U. As σ increases, z_σ^U declines, that is, $z_{\sigma\sigma}^U < 0$. This is because $z_{\sigma\sigma}^U = \tilde{b}_\sigma^U(\eta_\sigma^U + \eta^U \tilde{b}_\sigma^U / \tilde{b}^U) + \tilde{b}^U \partial(\eta^U + \eta^U \tilde{b}_\sigma^U / \tilde{b}^U) / \partial\sigma$, where based on what we have just said, the last derivative is negative. Then, when $\eta_\sigma^U + \eta^U \tilde{b}_\sigma^U / \tilde{b}^U = 0$, that is $z_\sigma^U = 0$, $z_{\sigma\sigma}^U < 0$, which in turn means that z^U reaches a maximum when $z_\sigma^U = 0$, and it then starts declining. This explains the shape of our curves in [the figure]. We have discussed in the text, and it is easy to see that when $\sigma = \lambda = \mu = 0$, then $\Pi^U = \Pi^P$. Then, as Π^U increases faster than Π^P as σ increases, a higher σ, with $\lambda = \mu = 0$, implies $\Pi^U > \Pi^P$. This establishes that a switch can take place if λ or μ are sufficiently small. Finally, differentiate $z^U - z^P$ with respect to λ or μ at $\sigma = \sigma^*$. We know that $z_\sigma^U - z_\sigma^P > 0$, and therefore the sign of σ^*_λ or σ^*_μ is the opposite of the sign of $z_\lambda^U - z_\lambda^P$ or $z_\mu^U - z_\mu^P$, which are both negative because λ and μ affect these expressions only through η^U. As a result σ^*

increases with λ or μ, and if λ or μ are too high the switch does not take place. QED

Appendix B: Proof of the Welfare Theorem

The strategy to prove this theorem is to show, first, that at σ^*, $W^U - W^P \geq 0$, and then that $W_\sigma^U - W_\sigma^P \geq 0$. Under the conditions of the choice-of-mode theorem, $\Pi_\sigma^U \geq \Pi_\sigma^P$. This means that at σ^*, when the firms switch from the P- to the U-mode, welfare is higher in the U-mode, and for larger σ, welfare does not switch back to the P-mode.

To show that $W^U - W^P \geq 0$, evaluate $W^U - W^P = N(\Pi^U - \Pi^P) + (1 - \sigma)$ $[\frac{1}{2}(1 - p^U + (1 - \mu)b^U)^2 + \mu b^U + \mu' h^U - \frac{1}{2}(1 - p^P + b^P)^2 - \mu' h^P] + \sigma[\frac{1}{2}(1 - p^U + (1 - \lambda)b^U)^2 + \lambda b^U + h^U - \frac{1}{2}(1 - p^P + b^P)^2 - h^P]$ at σ^* where $\Pi^U - \Pi^P = 0$ and $z^U y^U = z^P y^P$, which implies $p^U = p^P$ and $z^U = z^P$. The latter equality implies $y^U = y^P$ and therefore $\eta^U b^U = b^P$. We can rewrite $W^U - W^P$ at σ^* using all this information, suppressing for simplicity the superscript U, and rearranging terms, $W^U - W^P = (1 - \sigma)\frac{1}{2}[(1 - p + (1 - \mu)b)(1 - p + b) + \mu b(p - (1 - \mu)b)] + \sigma \frac{1}{2}[(1 - p + (1 - \lambda)b)(1 - p + b) + \lambda b(p - (1 - \lambda)b)] + \frac{1}{2}(1 - \eta)b - \frac{1}{2}(1 - p + \eta b)^2 + [(1 - \sigma)\mu' + \sigma](h^U - h^P)$. The terms in the first two square brackets are the number of users who buy times their surplus plus the number of users who do not buy times their surplus. This also explains why $0 \leq p - (1 - \mu)b \leq 1$ and $0 \leq p - (1 - \lambda)b \leq 1$. Beyond these boundaries the surplus of the non-innovating user is $\frac{1}{2} + \mu' b$ or $(\mu + \mu')b$, and $\frac{1}{2} + h$ or $\lambda b + h$ for the innovating users. As a result, $\mu b(p - (1 - \mu)b)$, $\lambda b(p - (1 - \lambda)b) \geq 0$ and it is easy to see that $h^U - h^P \geq 0$. Sum the first terms in the first two square brackets, weighed respectively by $(1 - \sigma)$ and σ, and subtract $\frac{1}{2}(1 - p + \eta b)^2$. This yields $\frac{1}{2}[(1 - p + b)(1 - p + \eta b) - (1 - p + \eta b)(1 - p + \eta b)] \geq 0$ because $\eta \leq 1$ and $1 - p + \eta b \geq 0$ because η is a weighted average between $(1 - \mu)$ and $(1 - \lambda)$, and $(p - (1 - \mu)b) \geq 0$. Since all the other terms in the expression for $W^U - W^P$ are non-negative, this establishes that at σ^*, $W^U \geq W^P$.

The next step is to show that $W_\sigma^U \geq W_\sigma^P$. The expression for W^P is (10) using the specific expressions for PS, CS^{nui} and CS^{ui} with $\lambda = \mu = 0$ and p, b and h computed for the P-mode, which means that $x = 0$ and η, $\tilde{b}$ and y are obtained from the problem of the firm under the P-mode. It is easy to see that in this case σ does not affect η, $\tilde{b}$ and y and therefore $W_\sigma^P = (1 - \mu')h^P$. For the U-mode, $W_\sigma^U = N\Pi_\sigma^U + (1 - \sigma)[(1 - p + (1 - \mu)b)$

$(-p_\sigma + (1 - \mu)b_\sigma) + \mu b_\sigma + \mu' h_\sigma^U] + \sigma[(1 - p + (1 - \lambda)b)(-p_\sigma + (1 - \lambda)b_\sigma) + \lambda b_\sigma + h_\sigma^U] + [(1 - p + (1 - \lambda)b)^2 - (1 - p + (1 - \mu) b)^2]/2 + (\lambda - \mu)b + (1 - \mu')h^U$, where apart from Π_σ^U and h^U we suppressed all the superscripts U. If $0 < \beta < \frac{1}{2}$ and σ is close to zero, $\Pi_\sigma^U \geq 0$. Moreover, $h^U - h^P \geq 0$. Thus, to establish the sign of $W_\sigma^U - W_\sigma^P$ we need to show that all the other terms of the expression for W_σ^U are non-negative. Start with the last term. Rewrite the difference of squares as the product of the sum and difference of the two terms, and collect $(\lambda - \mu)b$. We obtain $(\lambda - \mu)b(p - (1 - (\lambda + \mu)/2)b) \geq 0$ because $\lambda \geq \mu$ and $(p - (1 - (\lambda + \mu)/2)b) = \frac{1}{2}(p - (1 - \lambda)b + p - (1 - \mu)b)$ and we already established that $(p - (1 - \mu)b)$, $\lambda b(p - (1 - \lambda)b) \geq 0$. Finally, $(1 - p + (1 - \mu)b)(-p_\sigma + (1 - \mu)b_\sigma) + \mu b_\sigma = (1 - p + (1 - \mu)b)(-p_\sigma + b_\sigma) + \mu b_\sigma(p - (1 - \mu)b)$. We know that $(1 - p + (1 - \mu)b) \geq 0$, $(p - (1 - \mu)b) \geq 0$, and $-p_\sigma + b_\sigma = -(\eta_\sigma b + \eta b_\sigma)/(N + 1) + b_\sigma = -\eta_\sigma b/(N + 1) + b_\sigma[1 - \eta/(N + 1)] \geq 0$. This is because $\eta_\sigma \leq 0$, $b_\sigma = \tilde{b}_\sigma y + \tilde{b} y_\sigma \geq 0$, and $1 - \eta/(N + 1) > 0$ because $\eta \leq 1$. We obtain a similar result for the analogous term in λ. This establishes that $W_\sigma^U \geq W_\sigma^P$. QED

Appendix C: Proof of the Policy Theorem

Like in the previous theorem, the strategy to prove this theorem hinges on the fact that, as shown in the previous theorem, at $\sigma^* W^U - W^P \geq 0$, and then we study how $\Pi^U - \Pi^P$ and $W^U - W^P$ vary as we change γ or ξ. The logic is to check whether, under the conditions of the choice-of-mode theorem, changes in $\Pi^U - \Pi^P$ and $W^U - W^P$ go in the same direction.

In $W^U - W^P$ changes in γ do not affect any of the variables in the P-mode. They raise Π^U and h^U. The expression for W_γ^U is equivalent to W_σ^U in the proof of the previous theorem, without the last two terms and with subscripts γ instead of σ. Thus, to show that $W_\gamma^U - W_\gamma^P \geq 0$, we need to show that the second and third terms are non-negative. Analogously to the proof of the previous theorem, the second term can be written as $(1 - p + (1 - \mu)b)(-p_\gamma + b_\gamma) + \mu b_\gamma (p - (1 - \mu)b)$. We know that $(1 - p + (1 - \mu)b) \geq 0$, $(p - (1 - \mu)b) \geq 0$, and $-p_\gamma^U + b_\gamma^U = b_\gamma^U (1 - \eta^U/(N + 1)) \geq 0$. The same applies to the third term, which establishes that at σ^*, where $\Pi^U - \Pi^P = 0$ and $W^U - W^P \geq 0$, $W_\gamma^U + W_\gamma^P \geq 0$. This means that at σ^* increases in γ raise Π^U beyond Π^P, which induces firms to switch to the U-mode.

At the same time, welfare, which at σ^* is higher in the U-mode, cannot turn to be smaller than in the P-mode.

Consider now increases in ξ. We first show that if $\beta > \beta^*$ with $\beta^* < \frac{1}{2}$, $\Pi_\xi^U - \Pi_\xi^P \leq 0$. To see this, $z_\xi^U - z_\xi^P = \eta^U \Psi - 1$ where $\Psi \equiv [1 + (\tau/\xi)^\theta]^{(1-\theta)/\theta}$. If $0 < \theta < 1$, or $0 < \beta < \frac{1}{2}$, $\eta^U \leq 1$ but $\Psi \geq 1$. However, $\theta \to 0$ implies that Ψ becomes very large, and $\theta = 1$ implies $\Psi = 1$. Moreover, Ψ declines monotonically with θ. Consider $\partial \log\Psi/\partial\theta = -\theta^{-2}\log\Psi' + [(1 - \theta)/\theta] [(\tau/\xi)^\theta/\Psi']\log(\tau/\xi)$, where $\Psi' \equiv [1 + (\tau/\xi)^\theta]$. This expression is negative because we study cases in which $\tau/\xi < 1$. As a result, there is a threshold $\theta^* < 1$, or $\beta^* < \frac{1}{2}$, such that $\beta > \beta^* \to z_\xi^U - z_\xi^P < 0$, and vice versa. Thus, at $\sigma = \sigma^*$, increases in ξ induce a switch to the P-mode.

To check for $W_\xi^U + W_\xi^P$, using the logic of the proof of the previous theorem we can write $W_\xi^U = N\Pi_\xi + (1 - \sigma)[1 - p + (1 - \mu)b)(-p_\xi + b_\xi) + \mu b_\xi (p + (1 - \mu)b) + \mu' h_\xi] + \sigma[(1 - p + (1 - \lambda)b)(-p_\xi + b_\xi + \lambda b_\xi(p + (1 - \lambda)b) + h_\xi]$ where for simplicity we suppressed the superscripts U. The expression for W_ξ^P is the same with $\lambda = \mu = h_\xi = 0$ and the variables are all evaluated at the P-mode. Recall that, as noted in the proof of the previous theorem, at $\sigma = \sigma^*$, $p^U = p^P$ and $\eta^U b^U = b^P$. Then, in $W_\xi^U - W_\xi^P$, the difference between $(1 - p + (1 - \mu)b)(-p_\xi + b_\xi) + \mu b_\xi(p - (1 - \mu)b)$ and the equivalent term in W_ξ^P, weighed by $(1 - \sigma)$, and the difference between $(1 - p + (1 - \lambda)b)(-p_\xi + b_\xi + \lambda b_\xi(p - (1 - \lambda)b)$ and the equivalent term in W_ξ^P, weighed by σ, yields, after some algebra, $[(1 - \eta) + \sigma(1 - \sigma)(\lambda - \mu)^2 b]b_\xi \geq 0$. In addition, while h^P does not change with ξ, h^U increases in ξ. We conclude that at $\sigma = \sigma^*$, where $W^U \geq W^P$, the sign of $W_\xi^U - W_\xi^P$ is ambiguous and can very well be positive. Since $\beta > \beta^*$ implies $\Pi_\xi^U - \Pi_\xi^P \leq 0$, it may be that a higher ξ induces firms to switch to the P-mode while welfare is still higher under the U-mode. QED

Notes

1. The ratio of the marginal products of $b = (T^\beta + Y^\beta)^{1/\beta}$ with respect to T and Y, is equal to $(Y/T)^{1-\beta}$. With σ small, T is small and therefore Y/T is likely to be larger than 1. As a result, a lower β makes the impact of T on b higher relative to the impact of Y on b. Since a higher σ makes T higher and Y lower because the optimal s increases, the condition $0 < \beta < \frac{1}{2}$ says that the contribution of the higher T on b has to be strong enough to compensate by a sizable amount the negative effect on b due to a lower Y.

2. The intuition is that increases in ξ have a direct positive impact on Y both in the P- and U-mode. In addition, in the U-mode increases in ξ reduce s, which raises Y and reduces T. However, a higher β generates a more pronounced drop in s relative to $(1 - s)$ because when complementarity is strong, the increase in Y does not produce a strong decline in T due to the feedback produced by complementarity. As a result, when complementarity is weak, increases in ξ produce a stronger increase in b in the P- than in the U-mode.

References

Acemoglu, D., and J. Linn. 2004. Market size in innovation: Theory and evidence from the pharmaceutical industry. *Quarterly Journal of Economics* 119 (3): 1049–1090.

Adner, R., and R. Kapoor. 2010. Value creation in innovation ecosystems: How the structure of technological interdependence affects firm performance in new technology generations. *Strategic Management Journal* 31 (3): 306–333.

Afuah, A., and C. L. Tucci. 2012. Crowdsourcing as a solution to distant search. *Academy of Management Review* 37 (3): 355–375.

Agerfalk, P. J., and B. Fitzgerald. 2008. Outsourcing to an unknown workforce: exploring opensourcing as a global sourcing strategy. *Management Information Systems Quarterly* 32 (2): 385–410.

Akgün, A. E., H. Keskin, and J. C. Byrne. 2010. Procedural justice climate in new product development teams: Antecedents and consequences. *Journal of Product Innovation Management* 27 (7): 1096–1111.

Alchian, A. A., and H. Demsetz. 1972. Production, information costs, and economic organization. *American Economic Review* 62 (5): 777–795.

Allen, R. C. 1983. Collective invention. *Journal of Economic Behavior & Organization* 4 (1): 1–24.

Amabile, T. M., R. Conti, H. Coon, J. Lazenby, and M. Herron. 1996. Assessing the work environment for creativity. *Academy of Management Journal* 39 (5): 1154–1184.

Antorini, Y. M., A. M. J. Muñiz, and T. Askildsen. 2012. Collaborating with customer communities: Lessons from the Lego Group. *Sloan Management Review* 53 (3): 73–79.

Aoki, M. 2001. *Toward a Comparative Institutional Analysis*. MIT Press.

Arai, Y., and S. Kinukawa. 2014. Copyright infringement as user innovation. *Journal of Cultural Economics* 38 (2): 131–144.

Arora, A., W. M. Cohen, and J. P. Walsh. 2015. The Acquisition and Commercialization of Invention in American Manufacturing: Incidence and Impact. Working paper 20264, National Bureau of Economic Research (NBER), June 2014 (revised September 2015), Cambridge MA. Accessed January 27, 2016. http://www.nber.org/papers/w20264

Arora, A., A. Fosfuri, and A. Gambardella. 2001. Markets for technology and their implications for corporate strategy. *Industrial and Corporate Change* 10 (2): 419–451.

Arrow, K. J. 1962. Economic welfare and the allocation of resources for invention. In *The Rate and Direction of Inventive Activity: Economic and Social Factors*, ed. R. R. Nelson. Princeton University Press.

Arrow, K. J. 1974. *The Limits of Organization*. Norton.

Baker, W. E., and N. Bulkley. 2014. Paying it forward or rewarding reputation: Mechanisms of generalized reciprocity. *Organization Science* 25 (5): 1493–1510.

Baldwin, C. Y. 2008. Where do transactions come from? Modularity, transactions and the boundaries of firms. *Industrial and Corporate Change* 17 (1): 155–195.

Baldwin, C. Y. 2010. When Open Architecture Beats Closed: The Entrepreneurial Use of Architectural Knowledge. Working paper 10-063, Harvard Business School.

Baldwin, C. Y. 2015. Bottlenecks, Modules and Dynamic Architectural Capabilities. Finance working paper 15-028, Harvard Business School.

Baldwin, C. Y., and K. B. Clark. 2000. *Design Rules*, volume 1: *The Power of Modularity*. MIT Press.

Baldwin, C. Y., and K. B. Clark. 2006a. Between "knowledge" and the "economy": Notes on the scientific study of designs. In *Advancing Knowledge and the Knowledge Economy*, ed. Brian Kahin and Dominique Foray. MIT Press.

Baldwin, C. Y., and K. B. Clark. 2006b. The architecture of participation: Does code architecture mitigate free riding in the open source development model? *Management Science* 52 (7): 1116–1127.

Baldwin, C. Y., and J. Henkel. 2015. Modularity and intellectual property protection. *Strategic Management Journal* 36 (11): 1637–1655.

Baldwin, C. Y., and E. von Hippel. 2011. Modeling a paradigm shift: From producer innovation to user and open collaborative innovation. *Organization Science* 22 (6): 1399–1417.

Baldwin, C. Y., C. Hienerth, and E. von Hippel. 2006. How user innovations become commercial products: A theoretical investigation and case study. *Research Policy* 35 (9): 1291–1313.

Barnes, B., and D. R. Ulin. 1984. Liability for new products. *Journal of the American Water Works Association* 76 (2): 44–47.

Barnouw, E. 1966. *A Tower in Babel: A History of Broadcasting in the United States to 1933*. Oxford University Press.

Barrick, M. R., and K. M. Mount. 1991. The big five personality dimensions and job performance: A meta-analysis. *Personnel Psychology* 44 (1): 1–26.

Barrick, M. R., M. K. Mount, and T. A. Judge. 2001. Personality and performance at the beginning of the new millennium: What do we know and where do we go next? *International Journal of Selection and Assessment* 9 (1–2): 9–30.

Bator, F. M. 1958. The anatomy of market failure. *Quarterly Journal of Economics* 72 (3): 351–379.

Bauer, J., N. Franke, and P. Tuertscher. 2015. IP Norms in Online Communities: How User-Organized Intellectual Property Regulation Supports Innovation. Available at SSRN: http://ssrn.com/abstract=2718077.

Baumol, W. J. 2002. *The Free-Market Innovation Machine: Analyzing the Growth Miracle of Capitalism*. Princeton University Press.

Bayus, B. L. 2013. Crowdsourcing new product ideas over time: An analysis of the Dell IdeaStorm community. *Management Science* 59 (1): 226–244.

BEA (Bureau of Economic Analysis, U.S. Department of Commerce). 2016. Survey of Current Business Online 96, no. 1. Accessed January 31, 2016. http://www.bea.gov/scb/pdf/2015/12%20December/1215_gdp_and_the_economy.pdf

Benkler, Y. 2002. Coase's penguin, or, Linux and "the nature of the firm." *Yale Law Journal* 112 (3): 369–447.

Benkler, Y. 2004. Sharing nicely: On shareable goods and the emergence of sharing as a modality of economic production. *Yale Law Journal* 114 (2): 273–358.

Benkler, Y. 2006. *The Wealth of Networks: How Social Production Transforms Markets and Freedom*. Yale University Press.

Benkler, Y. 2016. When von Hippel innovation met the networked environment: Recognizing decentralized innovation. In *Revolutionizing Innovation: Users, Communities, and Open Innovation*, ed. Dietmar Harhoff and Karim R. Lakhani. MIT Press.

Bessen, J., and E. Maskin. 2009. Sequential innovation, patents, and imitation. *RAND Journal of Economics* 40 (4): 611–635.

Bin, G. 2013. A reasoned action perspective of user innovation: Model and empirical test. *Industrial Marketing Management* 42 (4): 608–619.

Blaxill, M., and R. Eckhardt. 2009. *The Invisible Edge: Taking Your Strategy to the Next Level Using Intellectual Property.* Portfolio.

Boudreau, K. J., and L. B. Jeppesen. 2015. Unpaid crowd complementors: The platform network effect mirage. *Strategic Management Journal* 36 (12): 1761–1777.

Boudreau, K. J., N. Lacetera, and K. R. Lakhani. 2011. Incentives and problem uncertainty in innovation contests: An empirical analysis. *Management Science* 57 (5): 843–863.

Boyle, J. 1997. A politics of intellectual property: Environmentalism for the Net? *Duke Law Journal* 47 (1): 87–116.

Braun, V., and C. Herstatt. 2008. The freedom fighters: How incumbent corporations are attempting to control user-innovation. *International Journal of Innovation Management* 12 (3): 543–572.

Braun, V., and C. Herstatt. 2009. *User-Innovation: Barriers to Democratization and IP Licensing.* Routledge.

Brynjolfsson, E., and J. H. Oh. 2012. The attention economy: Measuring the value of free goods on the Internet. Paper presented at the 33rd International Conference on Information Systems, 2012 Proceedings, Orlando. Accessed January 29, 2016. http://aisel.aisnet.org/icis2012/proceedings/EconomicsValue/9/

Buenstorf, G. 2003. Designing clunkers: Demand-side innovation and the early history of the mountain bike. In *Change, Transformation and Development*, ed. J. S. Metcalfe and U. Cantner. Springer.

Burda, M. C., D. S. Hamermesh, and P. Weil. 2007. Total Work, Gender and Social Norms. Discussion paper 2705, Institute for the Study of Labor, Bonn, Germany.

Bush, V. 1945. *Science: The Endless Frontier. A Report to the President by Vannevar Bush, Director of the Office of Scientific Research and Development, July 1945.* United States Government Printing Office. Accessed May 16, 2015. http://www.nsf.gov/about/history/vbush1945.htm

Casadesus-Masanell, R., and P. Ghemawat. 2006. Dynamic mixed duopoly: A model motivated by Linux vs. Windows. *Management Science* 52 (7): 1072–1084.

Castle Smurfenstein. 2016. "Official" Castle Smurfenstein Home Page. Accessed January 25, 2015. https://www.evl.uic.edu/aej/smurf.html

Chafee, Z., Jr. 1919. Freedom of speech in war time. *Harvard Law Review* 32 (8): 932–973.

Chamberlain, E. H. 1962. *The Theory of Monopolistic Competition: A Reorientation of the Theory of Value*, eighth edition. Harvard University Press.

Chandler, A. D., Jr. 1977. *The Visible Hand: The Managerial Revolution in American Business*. Harvard University Press.

Chesbrough, H. W. 2003. *Open Innovation: The New Imperative for Creating and Profiting from Technology*. Harvard Business School Press.

Clean Water Act. 1972. Federal Water Pollution Control Act Amendments of 1972, Public Law 92–500, *U.S. Statutes at Large* 86 (1972): 1251–1387, codified as amended at 33 U.S.C. § 1251 et seq., 1972.

Colombo, M. G., E. Piva, and C. Rossi-Lamastra. 2013. Authorising employees to collaborate with communities during working hours: When is it valuable for firms? *Long Range Planning* 46 (3): 236–257.

Committee for Orphan Medicinal Products and European Medicines Agency Scientific Secretariat. 2011. European regulation on orphan medicinal products: 10 years of experience and future perspectives. *Nature Reviews Drug Discovery* 10 (5): 341–349.

Constitution of the United States of America, As Amended. 2007. United States Government Printing Office.

Cooley, T. M. 1879. *A Treatise on the Law of Torts, or the Wrongs Which Arise Independently of Contract*. Callaghan.

Cooper, S., F. Khatib, A. Treuille, J. Barbero, J. Lee, M. Beenen, A. Leaver-Fay, D. Baker, Z. Popovic, and [57,000] Foldit players. 2010. Predicting protein structures with a multiplayer online game. *Nature* 466 (5): 756–760.

Costa, P. T., and R. R. McCrae. 1988. Personality in adulthood: A six-year longitudinal study of self-reports and spouse ratings on the NEO Personality Inventory. *Journal of Personality and Social Psychology* 54 (5): 853–863.

Costa, P. T., and R. R. McCrae. 1992. *Revised Neo Personality Inventory (NEO-PI-R) and NEO Five-Factor Inventory (NEO-FFI)*. Orlando: Psychological Assessment Resources.

Costa, P. T., and R. R. McCrae. 1995. Solid ground in the wetlands of personality: A reply to Block. *Psychological Bulletin* 117 (2): 216–220.

Cova, B., and T. White. 2010. Counter-brand and alter-brand communities: The impact of Web 2.0 on trial marketing approaches. *Journal of Marketing Management* 26 (3–4): 256–270.

Cowen, T., ed. 1988. *Public Goods and Market Failures: A Critical Examination.* George Mason University Press.

Crespi, G., C. Criscuolo, J. Haskel, and D. Hawkes. 2006. Measuring and understanding productivity in UK market services. *Oxford Review of Economic Policy* 22 (4): 560–572.

Dahl, T. E., and G. J. Allord. 1997. History of Wetlands in the Coterminous United States. Paper 2425, U.S. Geological Survey Water Supply. Accessed January 31, 2016. http://water.usgs.gov/nwsum/WSP2425/history.html

Dahl, D. W., C. Fuchs, and M. Schreier. 2015. Why and when consumers prefer products of user-driven firms: A social identification account. *Management Science* 61 (8): 1978–1988.

Dahlander, L. 2007. Penguin in a new suit: A tale of how *de novo* entrants emerged to harness free and open source software communities. *Industrial and Corporate Change* 16 (5): 913–943.

Dahlander, L., and M. W. Wallin. 2006. A man on the inside: Unlocking communities as complementary assets. *Research Policy* 35 (8): 1243–1259.

de Bruijn, E. 2010. On the viability of the Open Source Development model for the design of physical objects: Lessons learned from the RepRap project. Master of Science thesis, Tilburg University, Netherlands.

de Jong, J. P. J. 2013. User innovation by Canadian consumers: Analysis of a sample of 2,021 respondents. Unpublished paper commissioned by Industry Canada.

de Jong, J. P. J. 2015. Private communication with author.

de Jong, J. P. J. 2016, forthcoming. Surveying innovation in samples of individual end consumers. *European Journal of Innovation Management.* Available at SSRN: http://ssrn.com/abstract=2089422

de Jong, J. P. J., and E. de Bruijn. 2013. Innovation lessons from 3-D printing. *Sloan Management Review* 54 (2): 42–52.

de Jong, J. P. J., E. von Hippel, F. Gault, J. Kuusisto, and C. Raasch. 2015. Market failure in the diffusion of consumer-developed innovations: Patterns in Finland. *Research Policy* 44 (10): 1856–1865.

Delfanti, A. 2012. Tweaking genes in your garage: Biohacking between activism and entrepreneurship. In *Activist Media and Biopolitics: Critical Media*

Interventions in the Age of Biopower, ed. Wolfgang Sützl and Theo Hug. Innsbruck University Press.

DeMonaco, H., A. Ali, and E. von Hippel. 2006. The major role of clinicians in the discovery of off-label drug therapies. *Pharmacotherapy* 26 (3): 323–332.

Demsetz, H. 1988. The theory of the firm revisited. *Journal of Law, Economics, & Organization* 4 (1): 141–161.

Di Gangi, P. M., and M. Wasko. 2009. Steal my idea! Organizational adoption of free innovations from a free innovation community: A case study of Dell IdeaStorm. *Decision Support Systems* 48 (1): 303–312.

DMCA. 1998. *Digital Millennium Copyright Act of 1998*, Public Law 105-304. *United States Statutes at Large* 112: 2860.

Dosi, G., and R. R. Nelson. 2010. Technical change and industrial dynamics as evolutionary processes. In *Handbook of the Economics of Innovation*, volume 1, ed. Bronwyn H. Hall and Nathan Rosenberg. North-Holland.

Dosi, G., L. Marengo, and C. Pasquali. 2006. How much should society fuel the greed of innovators? On the relations between appropriability, opportunities and rates of innovation. *Research Policy* 35 (8): 1110–1121.

DoubleBlinded. 2016. DoubleBlinded: Placebo-controlled experiment kits for your supplements. Accessed March 30, 2016. http://doubleblinded.com/

Economides, N., and E. Katsamakas. 2006. Two-sided competition of proprietary vs. open source technology platforms and the implications for the software industry. *Management Science* 52 (7): 1057–1071.

Edwards, K. 1990. The interplay of affect and cognition in attitude formation and change. *Journal of Personality and Social Psychology* 59 (2): 202–216.

Electronic Frontier Foundation. 2013. Unintended consequences: Fifteen years under the DMCA. Accessed March 15, 2016. https://www.eff.org/pages/unintended-consequences-fifteen-years-under-dmca

Ensley, M. D., and K. M. Hmieleski. 2005. A comparative study of new venture top management team composition, dynamics and performance between university-based and independent start-ups. *Research Policy* 34 (7): 1091–1105.

Executive Order 12,291 of February 17, 1981. Federal Regulation. *Federal Register* 46 (33): 13193–13198. Accessed January 29, 2016. http://www.archives.gov/federal-register/codification/executive-order/12291.html

Executive Order 13563 of January 18, 2011. Improving Regulation and Regulatory Review. *Federal Register* 76 (14): 3821–3823. https://www.gpo.gov/fdsys/pkg/FR-2011-01-21/pdf/2011-1385.pdf. Accessed January 31, 2016.

Fama, E. F., and M. C. Jensen. 1983a. Separation of ownership and control. *Journal of Law & Economics* 26 (2): 301–325.

Fama, E. F., and M. C. Jensen. 1983b. Agency problems and residual claims. *Journal of Law & Economics* 26 (2): 327–349.

Fauchart, E., and M. Gruber. 2011. Darwinians, communitarians, and missionaries: The role of founder identity in entrepreneurship. *Academy of Management Journal* 54 (5): 935–957.

Fauchart, E., and E. von Hippel. 2008. Norms-based intellectual property systems: The case of French chefs. *Organization Science* 19 (2): 187–201.

Faullant, R., J. Füller, and K. Hutter. 2013. Fair play: Perceived fairness in crowdsourcing communities and its behavioral consequences. *Academy of Management Proceedings* 2013, no.1, Meeting Abstract Supplement 15433.

FCC. 2015. White Space Data Administration. Accessed December 14, 2015. http://www.fcc.gov/topic/white-space

Feist, G. J. 1998. A meta-analysis of personality in scientific and artistic creativity. *Personality and Social Psychology Review* 2 (4): 290–309.

Fisher, W. W., III. 2010. The implications for law of user innovation. *Minnesota Law Review* 94 (May): 1417–1477.

Fitzsimmons, J. A., and M. J. Fitzsimmons. 2001. *Service Management: Operations, Strategy, and Information Technology.* McGraw-Hill.

Franke, N., and F. Piller. 2004. Value creation by toolkits for user innovation and design: The case of the watch market. *Journal of Product Innovation Management* 21 (6): 401–415.

Franke, N., and M. Schreier. 2010. Why customers value self-designed products: The importance of process effort and enjoyment. *Journal of Product Innovation Management* 27 (7): 1020–1031.

Franke, N., and S. Shah. 2003. How communities support innovative activities: An exploration of assistance and sharing among end-users. *Research Policy* 32 (1): 157–178.

Franke, N., and E. von Hippel. 2003. Satisfying heterogeneous user needs via innovation toolkits: The case of Apache security software. *Research Policy* 32 (7): 1199–1215.

Franke, N., P. Keinz, and K. Klausberger. 2013. "Does this sound like a fair deal?" Antecedents and consequences of fairness expectations in the individual's decision to participate in firm innovation. *Organization Science* 24 (5): 1495–1516.

Franke, N., H. Reisinger, and D. Hoppe. 2009. Remaining within-cluster variance: A meta-analysis of the "dark side of clustering methods." *Journal of Marketing Management* 25 (3–4): 273–293.

Franke, N., E. von Hippel, and M. Schreier. 2006. Finding commercially attractive user innovations: A test of lead-user theory. *Journal of Product Innovation Management* 23 (4): 301–315.

Franklin, B. 1793; 2008. *The Autobiography of Benjamin Franklin 1706–1757.* Applewood Books. Originally published 1793.

Frey, K., C. Lüthje, and S. Haag. 2011. Whom should firms attract to open innovation platforms? The role of knowledge diversity and motivation. *Long Range Planning* 44 (5–6): 397–420.

Fuchs, C., and M. Schreier. 2011. Customer empowerment in new product development. *Journal of Product Innovation Management* 28 (1): 7–32.

Fuchs, C., E. Prandelli, and M. Schreier. 2010. The psychological effects of empowerment on consumer product demand. *Journal of Marketing* 74 (1): 65–79.

Füller, J. 2010. Refining virtual co-creation from a consumer perspective. *California Management Review* 52 (2): 98–122.

Füller, J., K. Hutter, and R. Faullant. 2011. Why co-creation experience matters? Creative experience and its impact on the quantity and quality of creative contributions. *R & D Management* 41 (3): 259–273.

Füller, J., R. Schroll, and E. von Hippel. 2013. User generated brands and their contribution to the diffusion of user innovations. *Research Policy* 42 (6–7): 1197–1209.

Fullerton, T. 2008. *Game Design Workshop: A Playcentric Approach to Creating Innovative Games.* Morgan Kaufmann.

Gallini, N., and S. Scotchmer. 2002. Intellectual property: When is it the best incentive system? In *Innovation Policy and the Economy*, volume 2, ed. Adam B. Jaffe, Josh Lerner, and Scott Stern. MIT Press.

Gambardella, A., C. Raasch, and E. von Hippel. 2016, forthcoming. The user innovation paradigm: impacts on markets and welfare. *Management Science.*

Gault, F. 2012. User innovation and the market. *Science & Public Policy* 39 (1): 118–128.

Gault, F. 2015. Measuring Innovation in All Sectors of the Economy. Working paper 2015-038, United Nations University and MERIT, Maastricht.

Gee, J. P. 2003. *What Video Games Have to Teach Us About Learning and Literacy*. Palgrave MacMillan.

George, J. M., and J. Zhou. 2001. When openness to experience and conscientiousness are related to creative behavior: An interactional approach. *Journal of Applied Psychology* 86 (3): 513–524.

Ghosh, R. A. 1998. Cooking pot markets: An economic model for the free trade of goods and services on the internet. *First Monday* 3 (3). Accessed January 15, 2016. http://firstmonday.org/ojs/index.php/fm/article/viewArticle/1516

Gobet, F., and H. A. Simon. 1998. Expert chess memory: Revisiting the chunking hypothesis. *Memory* 6 (3): 225–255.

Godin, B. 2006. The linear model of innovation: The historical construction of an analytical framework. *Science, Technology & Human Values* 31 (6): 639–667.

Goldberg, L. R. 1993. The structure of phenotypic personality traits. *American Psychologist* 48 (1): 26–34.

Goodman, L. A. 1961. Snowball sampling. *Annals of Mathematical Statistics* 32 (1): 117–151.

Goodman, P. S., R. Devadas, and T. L. Griffith Hughson. 1988. Analyzing the effectiveness of self-managing teams. In *Productivity in Organizations: New Perspectives from Industrial and Organizational Psychology*, ed. John P. Campbell and Richard J. Campbell. Jossey-Bass.

Green, P. E. 1977. A new approach to market segmentation. *Business Horizons* 20 (1): 61–73.

Greenberg, A. 2013. Evasion is the most popular jailbreak ever: Nearly seven million iOS devices hacked in four days. *Fortune*, February 8. Accessed January 11, 2016. http://www.forbes.com/sites/andygreenberg/2013/02/08/evasi0n-is-the-most-popular-jailbreak-ever-nearly-seven-million-ios-devices-hacked-in-four-days/

Greif, A. 2006. *Institutions and the Path to the Modern Economy: Lessons from Medieval Trade*. Cambridge University Press.

Griggs, R. C., M. Batshaw, M. Dunkle, R. Gopal-Srivastava, E. Kaye, J. Krischer, T. Nguyen, K. Paulus, and P. A. Merkel. 2009. Clinical research for rare disease: Opportunities, challenges, and solutions. *Molecular Genetics and Metabolism* 96 (1): 20–26.

Guba, E. G., and Y. S. Lincoln. 1994. Competing paradigms in qualitative research. In *Handbook of qualitative research*, ed. Norman K. Denzin and Yvonne S. Lincoln. SAGE.

Habicht, H., P. Oliveira, and V. Shcherbatiuk. 2012. User innovators: When patients set out to help themselves and end up helping many. *Die Unternehmung: Swiss Journal of Business Research and Practice* 66 (3): 277–295.

Halbinger, M. 2016. The role of intrinsic and extrinsic motivation in entrepreneurial activity. Unpublished paper, Zicklin School of Business, Baruch College, CUNY.

Hall, B., and D. Harhoff. 2004. Post-grant reviews in the U.S. patent system – Design choices and expected impact. *Berkeley Technology Law Journal* 19 (3): 989–1015.

Harhoff, D. 1996. Strategic spillovers and incentives for research and development. *Management Science* 42 (6): 907–925.

Harhoff, D., and K. R. Lakhani, eds. 2016. *Revolutionizing Innovation: Users, Communities, and Open Innovation*. MIT Press.

Harhoff, D., and P. Mayrhofer. 2010. Managing user communities and hybrid innovation processes: Concepts and design implications. *Organizational Dynamics* 39 (2): 137–144.

Harhoff, D., J. Henkel, and E. von Hippel. 2003. Profiting from voluntary information spillovers: How users benefit by freely revealing their innovations. *Research Policy* 32 (10): 1753–1769.

Harris, R. 2012. A building code with room for innovation. *New York Times,* October 5, 2012. Accessed January 28, 2016. http://green.blogs.nytimes.com/2012/10/05/a-building-code-with-room-for-innovation/

Hars, A., and S. Ou. 2002. Working for free? Motivations for participating in open-source projects. *International Journal of Electronic Commerce* 6 (3): 25–39.

Hart, O. 1995. *Firms, Contracts, and Financial Structure*. Oxford University Press.

Hemenway, K., and T. Calishain. 2004. *Spidering Hacks: 100 Industrial-Strength Tips and Tools*. O'Reilly.

Henkel, J. 2009. Champions of revealing: The role of open source developers in commercial firms. *Industrial and Corporate Change* 18 (3): 435–471.

Henkel, J., and E. von Hippel. 2004. Welfare implications of user innovation. *Journal of Technology Transfer* 30 (1): 73–87.

Henkel, J., C. Y. Baldwin, and W. C. Shih. 2013. IP modularity: Profiting from innovation by aligning product architecture with intellectual property. *California Management Review* 55 (4): 65–82.

Hertel, G., S. Niedner, and S. Herrmann. 2003. Motivation of software developers in Open Source projects: an Internet-based survey of contributors to the Linux kernel. *Research Policy* 32 (7): 1159–1177.

Hienerth, C. 2006. The commercialization of user innovations: The development of the rodeo kayaking industry. *R & D Management* 36 (3): 273–294.

Hienerth, C. 2016. Technique innovation. In *Revolutionizing Innovation: Users, Communities, and Open Innovation*, ed. Dietmar Harhoff and Karim R. Lakhani. MIT Press.

Hienerth, C., C. Lettl, and P. Keinz. 2014. Synergies among producer firms, lead users, and user communities: The case of the Lego producer–user ecosystem. *Journal of Product Innovation Management* 31 (4): 848–866.

Hienerth, C., E. von Hippel, and M. B. Jensen. 2014. User community vs. producer innovation development efficiency: A first empirical study. *Research Policy* 43 (1): 190–201.

Hill, B. M., and A. Shaw. 2014. Consider the redirect: A missing dimension of Wikipedia research. In *OpenSym'14: Proceedings of The International Symposium on Open Collaboration*, August 27–29, Berlin. Accessed January 28, 2016. https://mako.cc/academic/hill_shaw-consider_the_redirect.pdf

Hounshell, D. A. 1984. *From the American System to Mass Production, 1800–1932: The Development of Manufacturing Technology in the United States*. Johns Hopkins University Press.

Howe, J. 2006. The rise of crowdsourcing. *Wired*, June 1, 2006. Accessed January 29, 2016. http://www.wired.com/2006/06/crowds/

Hyysalo, S. 2009. User innovation and everyday practices: Micro-innovation in sports industry development. *R & D Management* 39 (3): 247–258.

Hyysalo, S., and S. Usenyuk. 2015. The user dominated technology era: Dynamics of dispersed peer-innovation. *Research Policy* 44 (3): 560–576.

IBC. 2009. 2009 International Building Code, International Code Council, Section [A]104.11: Alternate Materials, Design and Methods of Construction. As referenced in Utah Administrative Code R156–56. Building Inspector and Factory Built Housing Act Rule. Accessed January 31, 2016. https://law.resource.org/pub/us/code/ibr/icc.ibc.2009.pdf and http://www.rules.utah.gov/publicat/code/r156/r156-56.htm

Ironmonger, D. 2000. Household production and the household economy. Research paper, University of Melbourne, Department of Economics. Accessed January 29, 2016. http://fbe.unimelb.edu.au/__data/assets/pdf_file/0009/805995/759.pdf

Jacobides, M. G. 2005. Industry change through vertical disintegration: How and why markets emerged in mortgage banking. *Academy of Management Journal* 48 (3): 465–498.

Jefferson, T. 1819. III.28 To Isaac H. Tiffany, Monticello, April 4, 1819 [Letter from Thomas Jefferson to Isaac H. Tiffany]. In *Jefferson: Political Writings*, ed. Joyce Appleby and Terence Ball. Cambridge University Press, 1999.

Jenkins, H. 2008. *Convergence Culture: Where Old and New Media Collide*. New York University Press.

Jenkins, H., S. Ford, and J. Green. 2013. *Spreadable Media: Creating Value and Meaning in a Networked Culture*. New York University Press.

Jensen, M. C., and W. H. Meckling. 1994. The nature of man. *Journal of Applied Corporate Finance* 7 (2): 4–19.

Jeppesen, L. B. 2004. Profiting from innovative user communities: How firms organize the production of user modifications in the computer games industry. Working paper WP-04, Department of Industrial Economics and Strategy, Copenhagen Business School.

Jeppesen, L. B., and L. Frederiksen. 2006. Why do users contribute to firm-hosted user communities? The case of computer-controlled music instruments. *Organization Science* 17 (1): 45–63.

Jeppesen, L. B., and K. R. Lakhani. 2010. Marginality and problem solving effectiveness in broadcast search. *Organization Science* 21 (5): 1016–1033.

Joshi, A., L. E. Davis, and P. W. Palmberg. 1975. Electron spectroscopy. In *Methods of Surface Analysis*, ed. A. W. Czanderna. Elsevier.

Judge, T. A., J. E. Bono, R. Ilies, and M. W. Gerhardt. 2002. Personality and leadership: A qualitative and quantitative review. *Journal of Applied Psychology* 87 (4): 765–781.

Keller, K. L. 1993. Conceptualizing, measuring, and managing customer-based brand equity. *Journal of Marketing* 57 (1): 1–22.

Kharpal, A. 2014. Ikea "crushes" blogger in trademark spat. Accessed January 29, 2016. http://www.cnbc.com/2014/06/19/ikea-crushes-blogger-in-trademark-spat.html

Kim, Y. 2015. Consumer user innovation in Korea: An international comparison and policy implications. *Asian Journal of Technology Innovation* 23 (1): 69–86.

King, A., and G. Verona. 2014. Kitchen confidential? Norms for the use of transferred knowledge in gourmet cuisine. *Strategic Management Journal* 35 (11): 1645–1670.

Kline, S. J., and N. Rosenberg. 1986. An overview of innovation. In *The Positive Sum Strategy: Harnessing Technology for Economic Growth*, ed. Ralph Landau and Nathan Rosenberg. National Academies Press.

Kogut, B., and A. Metiu. 2001. Open-source software development and distributed innovation. *Oxford Review of Economic Policy* 17 (2): 248–264.

Kohler, T., J. Füller, K. Matzler, and D. Stieger. 2011. Co-creation in virtual worlds: The design of the user experience. *Management Information Systems Quarterly* 35 (3): 773–788.

Kotler, P. T. 1997. *Marketing Management: Analysis, Planning, Implementation, Control*, ninth edition. Prentice-Hall.

Kristof, A. L. 1996. Person-organization fit: An integrative review of its conceptualizations, measurement, and implications. *Personnel Psychology* 49 (1): 1–49.

Krugman, P., and R. Wells. 2006. *Economics*. Worth.

Kuan, J. W. 2001. Open source software as consumer integration into production. Unpublished paper, Haas School of Business, University of California Berkeley. Accessed January 30, 2016. http://papers.com/sol3/papers.cfm?abstract_id=259648

Kuhn, T. S. 1962;1970. *The Structure of Scientific Revolutions*, second edition, enlarged. University of Chicago Press.

Kuusisto, J., M. Niemi, and F. Gault. 2014. User innovators and their influence on innovation activities of firms in Finland. Working paper 2014-003, United Nations University-MERIT, Maastricht.

Ladd, J. 1957. *The Structure of a Moral Code: A Philosophical Analysis of Ethical Discourse Applied to the Ethics of the Navaho Indians*. Harvard University Press.

Lader, D., S. Short, and J. Gershuny. 2006. *The Time Use Survey, 2005: How We Spend Our Time*. Office for National Statistics, London.

Lafontaine, F., and M. Slade. 2007. Vertical integration and firm boundaries: The evidence. *Journal of Economic Literature* 45 (3): 629–685.

Lakhani, K., and E. von Hippel. 2003. How open source software works: "free" user-to-user assistance. *Research Policy* 32 (6): 923–943.

Lakhani, K. R., and R. G. Wolf. 2005. Why hackers do what they do: Understanding motivation and effort in free/open source software projects. In

Perspectives on Free and Open Source Software, ed. Joseph Feller, Brian Fitzgerald, Scott A. Hissam, and Karim R. Lakhani. MIT Press.

Lakhani, K., L. B. Jeppesen, P. A. Lohse, and J. A. Panetta. 2007. The Value of Openness in Scientific Problem Solving. Working paper 07-050, Harvard Business School.

Langlois, R. N. 1986. Rationality, institutions and explanation. In *Economics as a Process: Essays in the New Institutional Economics*, ed. Richard N. Langlois. Cambridge University Press.

Larkin, J., J. McDermott, D. P. Simon, and H. A. Simon. 1980. Expert and novice performance in solving physics problems. *Science* 208 (4450): 1335–1342.

Lehner, O. M. 2013. Crowdfunding social ventures: A model and research agenda. *Venture Capital: An International Journal of Entrepreneurial Finance* 15 (4): 289–311.

LePine, J. A., and L. Van Dyne. 2001. Voice and cooperative behavior as contrasting forms of contextual performance: Evidence of differential relationships with big five personality characteristics and cognitive ability. *Journal of Applied Psychology* 86 (2): 326–336.

Lerner, J., and J. Tirole. 2002. Some simple economics of open source. *Journal of Industrial Economics* 50 (2): 197–234.

Lettl, C., C. Herstatt, and H. G. Gemuenden. 2006. Users' contributions to radical innovation: Evidence from four cases in the field of medical equipment technology. *R & D Management* 36 (3): 251–272.

Levy, S. 2010. *Hackers: Heroes of the Computer Revolution*. O'Reilly.

Lewis, D., and S. Liebrand. 2014. What is #DIYPS (Do-It-Yourself Pancreas System)? Accessed December 30, 2015. http://diyps.org/

Lilien, G. L., P. D. Morrison, K. Searls, M. Sonnack, and E. von Hippel. 2002. Performance assessment of the lead user idea-generation process for new product development. *Management Science* 48 (8): 1042–1059.

Lin, L. 2008. Impact of user skills and network effects on the competition between open source and proprietary software. *Electronic Commerce Research and Applications* 7 (1): 68–81.

Linebaugh, K. 2014. Citizen hackers tinker with medical devices. *Wall Street Journal*, September 26, 2014. Accessed December 15, 2015. http://www.wsj.com/articles/citizen-hackers-concoct-upgrades-for-medical-devices-1411762843.

Lounsbury, J. W., N. Foster, H. Patel, P. Carmody, L. W. Gibson, and D. R. Stairs. 2012. An investigation of the personality traits of scientists versus non-scientists and their relationship with career satisfaction. *R & D Management* 42 (1): 47–59.

Lucas, R. E., E. Diener, A. Grob, E. M. Suh, and L. Shao. 2000. Cross-cultural evidence for the fundamental features of extraversion. *Journal of Personality and Social Psychology* 79 (3): 452–468.

Lüthje, C., C. Herstatt, and E. von Hippel. 2005. User innovators and "local" information: The case of mountain biking. *Research Policy* 34 (6): 951–965.

MacCormack, A., J. Rusnak, and C. Y. Baldwin. 2006. Exploring the structure of complex software designs: An empirical study of open source and proprietary code. *Management Science* 52 (7): 1015–1030.

Machlup, F., and E. Penrose. 1950. The patent controversy in the nineteenth century. *Journal of Economic History* 10 (1): 1–29.

Manz, C. C., and H. P. Sims, Jr. 1987. Leading workers to lead themselves: The external leadership of self-managing work teams. *Administrative Science Quarterly* 32 (1): 106–129.

Marx, M., D. Strumsky, and L. Fleming. 2009. Mobility, skills, and the Michigan non-compete experiment. *Management Science* 55 (6): 875–889.

Mauss, M. 1966. *The Gift: Forms and Functions of Exchange in Archaic Societies.* Trans. Ian Cunnison. Cohen & West. Originally published in 1925 as Essai sur le don. Forme et raison de l'échange dans les sociétés archaïques in *L'Année Sociologique.*

McCrae, R. R., and P. T. Costa. 1987. Validation of the five-factor model of personality across instruments and observers. *Journal of Personality and Social Psychology* 52 (1): 81–90.

McCrae, R. R., and P. T. Costa. 1997. Personality trait structure as a human universal. *American Psychologist* 52 (5): 509–516.

McCrae, R. R., and O. O. John. 1992. An introduction to the five-factor model and its applications. *Journal of Personality* 60 (2): 175–215.

Merges, R. P., and R. R. Nelson. 1994. On limiting or encouraging rivalry in technical progress: The effect of patent scope decisions. *Journal of Economic Behavior & Organization* 25 (1): 1–24.

Meyer, P. B. 2012. Open technology and the early airplane industry. Paper presented at annual meeting of Economic History Association, Vancouver, BC. Accessed January 30, 2016. http://www.law.nyu.edu/sites/default/files/ECM_PRO_069779.pdf

Mollick, E. 2014. The dynamics of crowdfunding: An exploratory study. *Journal of Business Venturing* 29 (1): 1–16.

Morrison, P. D., J. H. Roberts, and D. F. Midgley. 2004. The nature of lead users and measurement of leading edge status. *Research Policy* 33 (2): 351–362.

Morrison, P. D., J. H. Roberts, and E. von Hippel. 2000. Determinants of user innovation and innovation sharing in a local market. *Management Science* 46 (12): 1513–1527.

Muchinsky, P. M., and C. J. Monahan. 1987. What is person-environment congruence? Supplementary versus complementary models of fit. *Journal of Vocational Behavior* 31 (3): 268–277.

Murray, F., and S. Stern. 2007. Do formal intellectual property rights hinder the free flow of scientific knowledge? An empirical test of the anti-commons hypothesis. *Journal of Economic Behavior & Organization* 63 (4): 648–687.

Murray, F., P. Aghion, M. Dewatripont, J. Kolev, and S. Stern. 2009. Of Mice and Academics: Examining the Effect of Openness on Innovation. Working paper 14819, National Bureau of Economic Research (NBER), Cambridge MA.

Nagelkerke, N. J. D. 1991. A note on a general definition of the coefficient of determination. *Biometrika* 78 (3): 691–692.

Nambisan, S., and R. A. Baron. 2009. Virtual customer environments: Testing a model of voluntary participation in value co-creation activities. *Journal of Product Innovation Management* 26 (4): 388–406.

National Federation of Independent Business v. Sebelius. 2012. 132 S. Ct 2566, 2577 (2012). Opinion of Roberts, C.J., 18. Accessed January 29, 2016. http://www.supremecourt.gov/opinions/11pdf/11-393c3a2.pdf

Nelson, R. R. 1959. The economics of invention: A survey of the literature. *Journal of Business* 32 (2): 101–127.

Netcraft.com. 2015. March 2015 Web server survey. Accessed March 29, 2016. http://news.netcraft.com/archives/2015/03/19/march-2015-web-server-survey.html

Nightscout project. 2016. *Nightscout*. Accessed January 14, 2016. http://www.nightscout.info

Nishikawa, H., M. Schreier, and S. Ogawa. 2013. User-generated versus designer-generated products: A performance assessment at Muji. *International Journal of Research in Marketing* 30 (2): 160–167.

OECD. 2009. Society at a Glance 2009: OECD Social Indicators. Accessed January 30, 2016. http://www.oecdbookshop.org/get-it.php?REF=5KZ99FKTLPTB&TYPE=browse

OECD. 2015. National Accounts at a Glance. Accessed January 30, 2016. http://
www.keepeek.com/Digital-Asset-Management/oecd/economics/national
-accounts-at-a-glance-2015/household-final-and-actual-consumption_na
_glance-2015-table8-en#page1

OECD Guidelines. 2013. Standard concepts, definitions and classifications
for household wealth statistics. In *OECD Guidelines for Micro Statistics on
Household Wealth*. OECD Publishing; http://www.oecd.org/statistics/OECD
-Guidelines-for-Micro-Statistics-on-Household-Wealth.pdf. Accessed January
30, 2016.

Ogawa, S. 1998. Does sticky information affect the locus of innovation?
Evidence from Japanese convenience-store industry. *Research Policy* 26 (7–8):
777–790.

Ogawa, S., and K. Pongtanalert. 2011. Visualizing Invisible Innovation Content: Evidence from Global Consumer Innovation Surveys. Available at SSRN:
http://papers.ssrn.com/sol3/papers.cfm?abstract_id=1876186

Ogawa, S., and K. Pongtanalert. 2013. Exploring characteristics and motives of
consumer innovators: Community innovators vs. independent innovators.
Research Technology Management 56 (3): 41–48.

Oliar, D., and C. J. Sprigman. 2008. There's no free laugh (anymore): The
emergence of intellectual property norms and the transformation of stand-up
comedy. *Virginia Law Review* 94 (8): 1789–1867.

Oliveira, P., and E. von Hippel. 2011. Users as service innovators: The case of
banking services. *Research Policy* 40 (6): 806–818.

Oliveira, P., L. Zejnilovic, H. Canhão, and E. A. von Hippel. 2015. Innovation
by patients with rare diseases and chronic needs. *Orphanet Journal of Rare
Diseases* 10 (April Suppl.1): 41.

O'Mahony, S. 2003. Guarding the commons: How open source contributors
protect their work. *Research Policy* 32 (7): 1179–1198.

O'Mahony, S. 2007. The governance of open source initiatives: What does it
mean to be community managed? *Journal of Management & Governance* 11 (2):
139–150.

O'Mahony, S., and F. Ferraro. 2007. The emergence of governance in an open
source community. *Academy of Management Journal* 50 (5): 1079–1106.

Oslo Manual. 2005. *Oslo Manual: Guidelines for Collecting and Interpreting Innovation Data*, third edition. 2005. Statistical Office of the European Communities,
Organisation for Economic Co-Operation and Development.

Outdoor Foundation. 2009. A Special Report on Paddlesports 2009: Kayaking, Canoeing, Rafting. Accessed January 20, 2016. http://www.outdoorfoundation .org/research.paddlesports.html

Outdoor Industry Foundation. 2006. The Active Outdoor Recreation Economy: A \$730 Billion Annual Contribution to the U.S. Economy. Accessed January 20, 2015. http://www.outdoorindustry.org/images/researchfiles/ RecEconomypublic.pdf?26

Owen, I. 2015. e-Nabling the Future: A Global Network of Passionate Volunteers Using 3D Printing to Give the World a "Helping Hand." Accessed October 11, 2015. http://enablingthefuture.org/

Ozinga, J. R. 1999. *Altruism*. Praeger.

Patient Innovation. 2016. Patient Innovation: Sharing solutions, improving life. Accessed January 25, 2016. https://patient-innovation.com/

Penning, C. 1998. *Bike History. Die Erfolgsstory des Mountainbikes*. Delius Klasing.

Penrose, E. T. 1951. *The Economics of the International Patent System*. Johns Hopkins University Press.

Perry-Smith, J. E. 2006. Social yet creative: The role of social relationships in facilitating individual creativity. *Academy of Management Journal* 49 (1): 85–101.

Pine, B. J., II. 1993. *Mass Customization: The New Frontier in Business Competition*. Harvard Business School Press.

Pitt, L. F., R. T. Watson, P. Berthon, D. Wynn, and G. Zinkhan. 2006. The penguin's window: Corporate brands from an open-source perspective. *Journal of the Academy of Marketing Science* 34 (2): 115–127.

Poetz, M. K., and M. Schreier. 2012. The value of crowdsourcing: Can users really compete with professionals in generating new product ideas? *Journal of Product Innovation Management* 29 (2): 245–256.

Pongtanalert, K., and S. Ogawa. 2015. Classifying user-innovators: An approach to utilize user-innovator asset. *Journal of Engineering and Technology Management* 37 (July-September): 32–39.

Prügl, R., and M. Schreier. 2006. Learning from leading-edge customers at *The Sims*: opening up the innovation process using toolkits. *R & D Management* 36 (3): 237–251.

Raasch, C., C. Herstatt, and P. Lock. 2008. The dynamics of user innovation: Drivers and impediments of innovation activities. *International Journal of Innovation Management* 12 (3): 377–398.

Raasch, C., and E. von Hippel. 2013. Innovation process benefits: The journey as reward. *Sloan Management Review* 55 (1): 33–39.

Ram, K. 2013. Git can facilitate greater reproducibility and increased transparency in science. *Source Code for Biology and Medicine* 8 (1): 1–8.

Ramsar Convention. 1975. 1971. Convention on wetlands of international importance especially as waterfowl habitat, Concluded at Ramsar, Iran, on 2 February 1971 (No. 14583). *United Nations Treaty Series* 996: 245–267.

Raymond, E. A. 1999. *The Cathedral and the Bazaar: Musings on Linux and Open Source by an Accidental Revolutionary*. O'Reilly.

Riggs, W., and E. von Hippel. 1994. The impact of scientific and commercial values on the sources of scientific instrument innovation. *Research Policy* 23 (4): 459–469.

Riggs, W. M., and M. J. Parker. 1975. Surface analysis by x-ray photoelectron spectroscopy. In *Methods of Surface Analysis*, ed. A. W. Czanderna. Elsevier.

Rivette, K. G., and D. Kline. 1999. *Rembrandts in the Attic: Unlocking the Hidden Value of Patents*. Harvard Business School Press.

Robinson, J. 1933. *The Economics of Imperfect Competition*. Macmillan.

Rodwell, C., and S. Aymé, eds. 2014. 2014 Report on the State of the Art of Rare Disease Activities in Europe. Accessed January 31, 2016. http://www.eucerd.eu/upload/file/Reports/2014ReportStateofArtRDActivities.pdf

Roin, B. N. 2013. Solving the Problem of New Uses. Working paper, MIT Sloan School of Management. Available at SSRN: http://ssrn.com/abstract=2337821

Romer, P.M. 1990. Endogenous technological change. *Journal of Political Economy* 98 (5): S71–S102.

Rothmann, S., and E. P. Coetzer. 2003. The big five personality dimensions and job performance. *South African Journal of Industrial Psychology* 29 (1): 68–74.

Sahlins, M. 1972. *Stone Age Economics*. Aldine de Gruyter.

Samuelson, P. 2015. Freedom to tinker. UC Berkeley Public Law Research paper 2605195, University of California, Berkeley. *Theoretical Inquiries in Law*, forthcoming. Available at SSRN: http://papers.ssrn.com/sol3/Papers.cfm?abstract_id=2605195

Sandvig, C. 2012. What are community networks an example of? A response. In *Connecting Canadians: Investigations in Community Informatics*, ed. Andrew Clement, Michael Gurstein, Graham Longford, Marita Moll, and Leslie Regan Shade. AU Press, Athabascau University.

Schaffer, C. M., and P. E. Green. 1998. Cluster-based market segmentation: Some further comparisons of alternative approaches. *Journal of the Market Research Society* 40 (2): 155–163.

Schell, J. 2008. *The Art of Game Design: A Book of Lenses*. Morgan Kaufmann.

Schilling, M. A. 2000. Toward a general modular systems theory and its application to interfirm product modularity. *Academy of Management Review* 25 (2): 312–334.

Schoen, S. D. 2003. EOF—Give TCPA an owner override. *Linux Journal* (116): 14.

Schreier, M., C. Fuchs, and D. W. Dahl. 2012. The innovation effect of user design: Exploring consumers' innovation perceptions of firms selling products designed by users. *Journal of Marketing* 76 (5): 18–32.

Schumpeter, J. A. 1934. *The Theory of Economic Development: An Inquiry into Profits, Capital, Credit, Interest, and the Business Cycle*. Harvard University Press. Originally published in German in 1912; first English translation published in 1934.

Schweisfurth, T. G., and C. Raasch. 2015. Embedded lead users: The benefits of employing users for corporate innovation. *Research Policy* 44 (1): 168–180.

Scott, W. R. 2001. *Institutions and Organizations: Ideas, Interests, Identities*. SAGE.

Scott, S. G., and R. A. Bruce. 1994. Determinants of innovative behavior: A path model of individual innovation in the workplace. *Academy of Management Journal* 37 (3): 580–607.

Sen, R. 2007. A strategic analysis of competition between open source and proprietary software. *Journal of Management Information Systems* 24 (1): 233–257.

Shah, S. 2000. Sources and Patterns of Innovation in a Consumer Products Field: Innovations in Sporting Equipment. Working paper 4105, MIT Sloan School of Management.

Shah, S. K., and M. Tripsas. 2007. The accidental entrepreneur: The emergent and collective process of user entrepreneurship. *Strategic Entrepreneurship Journal* 1 (1–2): 123–140.

Shapeways. Run your business on Shapeways with 3D printing. Accessed January 14, 2016. https://www.shapeways.com/sell

Sherwin, C. W., and R. S. Isenson. 1967. Project HINDSIGHT: A Defense Department study of the utility of research. *Science* 156 (3782): 1571–1577.

Shirky, C. 2010. *Cognitive Surplus: How Technology Makes Consumers into Collaborators*. Penguin.

Simon, H. A. 1981. *The Sciences of the Artificial*, second edition. MIT Press.

Singer, S., J. E. Amorós, and D. Moska. 2015. *Global Entrepreneurship Monitor: 2014 Global Report*. Global Entrepreneurship Research Association, London Business School.

Smith, A. 1776; 1976. *An Inquiry into the Nature and Causes of the Wealth of Nations*. University of Chicago Press. Originally published 1776; Edwin Cannan's edition originally published in 1904 by Methuen & Co.

Song, P., J. Gao, Y. Inagaki, N. Kukudo, and W. Tang. 2012. Rare diseases, orphan drugs, and their regulation in Asia: Current status and future perspectives. *Intractable & Rare Diseases Research* 1 (1): 3–9.

Stallman, R. M. 2002. *Free Software Free Society: Selected Essays of Richard Stallman*. GNU Press, Free Software Foundation.

Statistics Finland. 2016. Innovation 2014. Science, Technology and Information Society, Helsinki, March 24, 2016. Accessed April 18, 2016. http://www.stat.fi/til/inn/2014/inn_2014_2016-03-24_tie_001_en.html

Steam Workshop. 2016. *Steam Community: Steam Workshop*. Accessed January 15, 2016. http://steamcommunity.com/workshop/

Stern, S. 2004. Do scientists pay to be scientists? *Management Science* 50 (6): 835–853.

Stock, R. M., P. Oliveira, and E. von Hippel. 2015. Impacts of hedonic and utilitarian motives on the innovativeness of user-developed innovations. *Journal of Product Innovation Management* 32 (3): 389–403.

Stock, R. M., E. von Hippel, and N. L. Gillert. 2016. Impact of personality traits on consumer innovation success. *Research Policy* 45 (4): 757–769.

Stoltz, M. 2015. New "Breaking Down Barriers to Innovation Act" targets many of DMCA Section 1201's problems. Accessed April 18, 2016. https://www.eff.org/deeplinks/2015/04/new-breaking-down-barriers-innovation-act-targets-many-dmca-section-1201s-problems

Strandburg, K. J. 2008. Users as innovators: Implications for patent doctrine. *University of Colorado Law Review* 79 (2): 467–544.

Suh, N. P. 1990. *The Principles of Design*. Oxford University Press.

Sung, S. Y., and J. N. Choi. 2009. Do big five personality factors affect individual creativity? the moderating role of extrinsic motivation. *Social Behavior and Personality* 37 (7): 941–956.

Svensson, P. O., and R. K. Hartmann. 2016. Policies to Promote User Innovation: Evidence from Swedish Hospitals on the Effects of Access to Makerspaces on Innovation by Clinicians. Working paper, MIT Sloan School of Management. Available at SSRN: http://papers.ssrn.com/sol3/papers.cfm?abstract_id=2701983

Syam, N. B., and A. Pazgal. 2013. Co-creation with production externalities. *Marketing Science* 32 (5): 805–820.

Tadelis, S., and O. E. Williamson. 2013. Transaction cost economics. In *Handbook of Organizational Economics*, ed. Robert Gibbons and John Roberts. Princeton University Press.

Taft, S. L. 2001. *The River Chasers: A History of American Whitewater Paddling*. Flowing Water Press and Alpen Books.

Teece, D. J. 1986. Profiting from technological innovation: Implications for integration, collaboration, licensing and public policy. *Research Policy* 15 (6): 285–305.

Teece, D. J. 1996. Firm organization, industrial structure, and technological innovation. *Journal of Economic Behavior & Organization* 31 (2): 193–224.

Teece, D. J. 2000. *Managing Intellectual Capital: Organizational, Strategic, and Policy Dimensions*. Oxford University Press.

Teixeira, Joaquina. 2014. Balloons at different heights to encourage a child [with Angelman's syndrome] to get up and walk. Accessed on Patient-Innovation.com on January 29, 2016. https://patient-innovation.com/condition/angelmans-syndrome?post=466

Torrance, A. W. 2010. Synthesizing law for synthetic biology. *Minnesota Journal of Law, Science & Technology* 11 (2): 629–665.

Torrance, A. W. 2015. Private communication with author.

Torrance, A. W., and L. J. Kahl. 2014. Bringing standards to life: Synthetic biology standards and intellectual property. *Santa Clara High Technology Law Journal* 30 (2): 199–230.

Torrance, A. W., and E. von Hippel. 2015. The right to innovate. *Detroit College of Law at Michigan State University Law Review* (2): 793–829.

Tseng, M. M., and F. Piller, eds. 2003. *The Customer Centric Enterprise: Advances in Mass Customization and Personalization*. Springer.

Ulrich, K. T., and S. D. Eppinger. 2016. *Product Design and Development*, sixth edition. McGraw-Hill.

UN. 2002. United Nations, European Commission, International Monetary Fund, Organisation for Economic Cooperation and Development, United

Nations Conference on Trade and Development, and World Trade Organization. *Manual on Statistics of International Trade in Services.*

Urban, G. L., and J. R. Hauser. 1993. *Design and Marketing of New Products,* second edition. Prentice-Hall.

Urban, G. L., and E. von Hippel. 1988. Lead user analyses for the development of new industrial products. *Management Science* 34 (5): 569–582.

van der Boor, P., P. Oliveira, and F. Veloso. 2014. Users as innovators in developing countries: The sources of innovation and diffusion in mobile banking services. *Research Policy* 43 (9): 1594–1607.

Vargo, S. L., and R. F. Lusch. 2004. The four service marketing myths: Remnants of a goods-based, manufacturing model. *Journal of Service Research* 6 (4): 324–335.

Vissers, G., and B. Dankbaar. 2002. Creativity in multidisciplinary new product development teams. *Creativity and Innovation Management* 11 (1): 31–42.

Von Ahn, L., and L. Dabbish. 2008. Designing games with a purpose. *Communications of the ACM* 51 (8): 58–67.

von Hippel, E. 1982. Appropriability of innovation benefit as predictor of the source of innovation. *Research Policy* 11 (2): 95–115.

von Hippel, E. 1986. Lead users: A source of novel product concepts. *Management Science* 32 (7): 791–805.

von Hippel, E. 1988. *The Sources of Innovation.* Oxford University Press.

von Hippel, E. 1994. "Sticky information" and the locus of problem-solving: Implications for innovation. *Management Science* 40 (4): 429–439.

von Hippel, E. 2005. *Democratizing Innovation.* MIT Press.

von Hippel, E., and S. N. Finkelstein. 1979. Analysis of innovation in automated clinical chemistry analyzers. *Science and Public Policy* 6 (1): 24–37.

von Hippel, E., and R. Katz. 2002. Shifting innovation to users via toolkits. *Management Science* 48 (7): 821–833.

von Hippel, E., and G. von Krogh. 2003. Open source software and the "private-collective" innovation model: Issues for organization science. *Organization Science* 14 (2): 209–223.

von Hippel, E. A., J. P. J. de Jong, and S. Flowers. 2012. Comparing business and household sector innovation in consumer products: Findings from a representative survey in the United Kingdom. *Management Science* 58 (9): 1669–1681.

von Hippel, E. A., H. J. DeMonaco, and J. P. J. de Jong. 2016, forthcoming. Market failure in the diffusion of clinician-developed innovations: The case of off-label drug discoveries. *Science and Public Policy*. Available at SSRN: http://papers.ssrn.com/sol3/papers.cfm?abstract_id=2275562

von Hippel, E., S. Ogawa, and J. P. J. de Jong. 2011. The age of the consumer-innovator. *Sloan Management Review* 53 (1): 27–35.

von Hippel, W., L. E. Hayward, E. Baker, S. L. Dubbs, and E. von Hippel. 2016. Boredom as a spur to innovation. Working paper, University of Queensland, Brisbane.

von Krogh, G., S. Spaeth, and K. R. Lakhani. 2003. Community, joining, and specialization in open source software innovation: A case study. *Research Policy* 32 (7): 1217–1241.

Walsh, J. P., C. Cho, and W. M. Cohen. 2005. View from the bench: Patents and materials transfers. *Science* 309 (5743): 2002–2003.

Warncke-Wang, M., V. Ranjan, L. Terveen, and B. Hecht. 2015. Misalignment between supply and demand of quality content in peer production communities. In Proceedings of the Ninth International AAAI Conference on Web and Social Media. Accessed on January 31, 2016. http://www.aaai.org/ocs/index.php/ICWSM/ICWSM15/paper/view/10591

Warren, S. D., and L. D. Brandeis. 1890. The right to privacy. *Harvard Law Review* 4 (5): 193–220.

Watershed Protection Act. 1954. *Watershed Protection and Flood Prevention Act*, Public Law 83-566, *U.S. Statutes at Large* 68 (1954): 666.

Webb, D. J., C. L. Green, and T. G. Brashear. 2000. Development and validation of scales to measure attitudes influencing monetary donations to charitable organizations. *Journal of the Academy of Marketing Science* 28 (2): 299–309.

Welch, S. 1975. Sampling by referral in a dispersed population. *Public Opinion Quarterly* 39 (2): 237–245.

West, J., and K. R. Lakhani. 2008. Getting clear about communities in open innovation. *Industry and Innovation* 15 (2): 223–231.

Wicks, P., T. E. Vaughan, M. P. Massagli, and J. Heywood. 2011. Accelerated clinical discovery using self-reported patient data collected online and a patient-matching algorithm. *Nature Biotechnology* 29 (5): 411–414.

Williamson, J. M., J. W. Lounsbury, and L. D. Han. 2013. Key personality traits of engineers for innovation and technology development. *Journal of Engineering and Technology Management* 30 (2): 157–168.

Williamson, O. 1973. Markets and hierarchies: Some elementary consider-ations. *American Economic Review* 63 (2): 316–325.

Williamson, O. E. 1985. *The Economic Institutions of Capitalism*. Free Press.

Williamson, O. E. 2000. The new institutional economics: Taking stock, look-ing ahead. *Journal of Economic Literature* 38 (3): 595–613.

Willis, N. 2007. Lessons learned from open source Xara's failure. Linux.com: News for the Open Source Professional. Accessed June 12, 2016. https://www.linux.com/news/lessons-learned-open-source-xaras-failure

Winston Smith, S., and S. K. Shah. 2013. Do innovative users generate more useful insights? An analysis of corporate venture capital investments in the medical device industry. *Strategic Entrepreneurship Journal* 7 (2): 151–167.

Winter, S. G. 2010. The replication perspective on productive knowledge. In *Dynamics of Knowledge, Corporate Systems and Innovation*, ed. Hiroyuki Itami, Ken Kusunoki, Tsuyoshi Numagami, and Akira Takeishi. Springer.

Wolfradt, U., and J. E. Pretz. 2001. Individual differences in creativity: Personal-ity, story writing, and hobbies. *European Journal of Personality* 15 (4): 297–310.

Wörter, M., K. Trantopoulos, E. von Hippel, and G. von Krogh. 2016. The Performance Effects of User Innovations on Firms. Working paper, ETH Zurich.

Wunsch-Vincent, S., and G. Vickery. 2007. Participative web: User-created content. Report prepared for Working Party on the Information Economy, Organisation for Economic Co-Operation and Development, and Directorate for Science Technology and Industry. Accessed January 15, 2016. http://www.oecd.org/sti/38393115.pdf

Yee, N. 2006. Motivations for play in online games. *Cyberpsychology & Behavior* 9 (6): 772–775.

Zajonc, R.B. 1968. Attitudinal effects of mere exposure. *Journal of Personality and Social Psychology* 9 (2): 1–27.

Zeithaml, V., and M. J. Bitner. 2003. *Services Marketing: Integrating Customer Focus Across the Firm*, third edition. McGraw-Hill.

Zhao, H., and S. E. Seibert. 2006. The big five personality dimensions and entre-preneurial status: A meta-analytical review. *Journal of Applied Psychology* 91 (2): 259–271.

Zicherman, G., and C. Cunningham. 2011. *Gamification by Design: Implement-ing Game Mechanics in Web and Mobile Apps*. O'Reilly Media.

Index